AF739346

"Attention to wild animal suffering may be the most important recent development in animal ethics. It threatens to reverse a host of common judgments in environmental advocacy, conservation biology, and philanthropic cause prioritization. Kyle Johannsen has done an admirable job explaining why it matters, why we ought to do something about it, and feasible strategies for helping. This book is essential reading for anyone concerned about wild animals and our responsibilities to them."

**Bob Fischer**, Department of Philosophy,
Texas State University

"*Wild Animal Ethics* is an excellent book that makes a powerful case for reducing wild animal suffering. Johannsen convincingly shows that assisting wild animals should be a top moral and political priority whether we care about animal welfare, animal rights, or both. He also addresses a number of important issues ranging from the ethics of killing predators to the ethics of genetic modification. Every reader will benefit from engaging with the arguments in this book, and wild animals will benefit enormously if we accept the main conclusions."

**Jeff Sebo**, Affiliated Professor of Bioethics,
Medical Ethics, and Philosophy, New York University

"Much has been written about the appalling suffering that we inflict on animals through practices such as factory farming. But the suffering of animals in the wild is vastly more extensive. *Wild Animal Ethics* defends the claim that although we are not the cause of this pervasive suffering, we nevertheless have moral reasons to try to mitigate it. Johannsen's arguments are neither naïve nor utopian. They make a convincing case for the importance of research into ways of intervening beneficently in the natural world, particularly through techniques of genetic modification that could gradually, and without harmful side effects, diminish predation and modes of reproduction that doom the vast majority of offspring to early, painful deaths. The suffering of animals in the wild is a serious moral issue, to which this book is a sensible, well-argued, and humane response."

**Jeff McMahan**, White's Professor of Moral Philosophy,
University of Oxford

# WILD ANIMAL ETHICS

Though many ethicists have the intuition that we should leave nature alone, Kyle Johannsen argues that we have a duty to research safe ways of providing large-scale assistance to wild animals. Using concepts from moral and political philosophy to analyze the issue of wild animal suffering (WAS), Johannsen explores how a collective, institutional obligation to assist wild animals should be understood. He claims that with enough research, genetic editing may one day give us the power to safely intervene without perpetually interfering with wild animals' liberties.

Questions addressed include:

- In what way is nature valuable and is intervention compatible with that value?
- Is intervention a requirement of justice?
- What are the implications of WAS for animal rights advocacy?
- What types of intervention are promising?

Expertly moving the debate about human relations with wild animals beyond its traditional confines, *Wild Animal Ethics* is essential reading for students and scholars of political philosophy and political theory studying animal ethics, environmental ethics, and environmental philosophy.

**Kyle Johannsen** is Adjunct Assistant Professor in the Department of Philosophy at Queen's University, Kingston. He is primarily interested in social and political philosophy, and in animal and environmental ethics. His first monograph – *A Conceptual Investigation of Justice* was published with Routledge in 2018, and his work has appeared in journals such as *Dialogue*, *Environmental Values*, *Ethical Perspectives*, *Ethical Theory and Moral Practice*, and *Res Publica*.

# WILD ANIMAL ETHICS

## The Moral and Political Problem of Wild Animal Suffering

*Kyle Johannsen*

Routledge
Taylor & Francis Group
NEW YORK AND LONDON

First published 2021
by Routledge
52 Vanderbilt Avenue, New York, NY 10017

and by Routledge
2 Park Square, Milton Park, Abingdon, Oxon OX14 4RN

*Routledge is an imprint of the Taylor & Francis Group, an Informa business*

© 2021 Taylor & Francis

The right of Kyle Johannsen to be identified as author of this work has been asserted by him in accordance with sections 77 and 78 of the Copyright, Designs and Patents Act 1988.

All rights reserved. No part of this book may be reprinted or reproduced or utilized in any form or by any electronic, mechanical, or other means, now known or hereafter invented, including photocopying and recording, or in any information storage or retrieval system, without permission in writing from the publishers.

*Trademark notice*: Product or corporate names may be trademarks or registered trademarks, and are used only for identification and explanation without intent to infringe.

*Library of Congress Cataloging-in-Publication Data*
A catalog record for this title has been requested

ISBN: 978-0-367-27568-6 (hbk)
ISBN: 978-0-367-27570-9 (pbk)
ISBN: 978-0-429-29667-3 (ebk)

Typeset in Bembo
by Newgen Publishing UK

# CONTENTS

# ACKNOWLEDGMENTS

This book was significantly improved by the help of a number of different people. For written comments on draft chapters and/or relevant conference papers, I'd like to thank Peter Andes, Sue Donaldson, Oscar Horta, Will Kymlicka, Eze Paez, Eldon Soifer, and Brian Tomasik. I'd also like to thank the referees who reviewed my book proposal, one of whom I learned is John Hadley and the other of whom remains anonymous. Their comments helped to determine how my book would develop, as it was largely unwritten when I submitted my proposal.

Much of the material and ideas in this book were presented in earlier forms at various conferences, including those hosted by the Atlantic Region Philosophers' Association, the Canadian Philosophical Association, the Canadian Society for Environmental Philosophy, the International Association for Philosophy of Law and Social Philosophy, the North American Society for Social Philosophy, and Minding Animals International. Some of the ideas I discuss in my book were also presented at Dalhousie University's Philosophy Colloquium Series, at Trent University Durham's Annual Research Day, and at Brock University's Rethinking Canid–Human Relations Conference. Last but not least, a very early draft of what would become Chapter 4 was discussed at Queen's University's Justice League Research Group. I'm thankful to my audiences for their comments.

Most of the content in this book has not been previously published, but it does contain content from my 2017 article "Animal Rights and the Problem of r-Strategists," *Ethical Theory and Moral Practice* 20: 333–45; and from my 2020 article "To Assist or Not to Assist? Assessing the Potential Moral Costs of Humanitarian Intervention in Nature," *Environmental Values* 29: 29–45. Chapter 4, in particular, is basically a shortened version of the article I published in *Environmental Values*. I'm grateful that both *Ethical Theory and Moral Practice* and *Environmental Values* permit their authors to reproduce content.

Special thanks as well to my wife, Melissa Teodoro, for her love and support, and to the publishing team at Routledge, particularly Natalja Mortensen and Charlie Baker. The help and advice they provided are much appreciated.

# 1

# INTRODUCTION

## 1.1 The Subject of This Book

In recent years, animal rights theorists have increasingly come to appreciate the extent to which wild animals fail to live flourishing lives. Though we've always known that the wild is a nasty place where predators lethally attack prey, only recently have most animal rights theorists come to realize that the majority of wild animals fail to flourish. In fact, what we know about wild animal reproduction suggests that the majority of sentient beings born into the world may not even live lives worth living. After all, only some wild animals (K-strategists) protect their genes by restricting reproduction to a small number of cared-for offspring. Many animals protect their genes by producing large numbers of uncared for offspring. This evolutionary reproductive strategy, normally referred to as the r-strategy,[1] is used by many lizards, amphibians, fish, and small mammals. Instead of restricting reproduction and providing intensive care, r-strategists produce a large quantity of offspring, the majority of whom die from disease, starvation, injury, exposure, or predation shortly after birth. In some cases, the death that an r-strategist infant experiences is quick, albeit painful, for example, being quickly eaten by a predator. In other cases, it can be slow and grueling (dying from starvation or exposure takes a while). And even when their death happens quickly, it typically occurs before they've had the opportunity to become comfortable with or competent to navigate their environment. In other words, your typical r-strategist infant crawls around uncomfortably for a short while, after which it dies either a quick and painful death, or a slow and painful death. Only a very small number of r-strategist offspring live long enough to reach a point in their lives where they're competent to manage the dangers of their environment and able to reproduce.[2]

For example, consider the results of a population study concerning *Podarcis muralis*: the common wall lizard.[3] The population in the study (which was located in a cemetery in France) started at a total of 570 members (570 eggs). Of those 570, 194 lived to one year of age, 48 reached their second year of life, 25 made it to year three, and only 12 reached year four. Some of the eggs did not hatch at all, so a portion of the 'mortality' represented by the drop from 570 to 194 did not involve the death of sentient beings. Still, even if we (implausibly) assume that the initial drop in population size is entirely attributable to unhatched eggs; most of these wall lizards experienced an abysmal fate. Only 48 of 194 lizards reached the age of sexual maturity (year 2): that's roughly 25% of the population.[4] Considering that a healthy, well-nourished wall lizard can live for up to 10 years, premature death was apparently the norm for this population.[5]

The wall lizards in the above study are not idiosyncratic. Many amphibians and lizards have similar life histories. High early-life mortality rates are normal for crocodilians, turtles, and amphibians that develop indirectly (have a larval stage in their life history). In fact, species in these groups typically experience an early-life death rate of greater than 90%.[6]

In combination, the fact that many wild animal species are r-strategist, that r-strategists (by definition) have far higher reproduction rates than K-strategists, and that most r-strategist young live short painful lives, is concerning. In light of these facts, I believe the below argument is sound.

> Premise 1: A life that's filled with suffering and ends shortly after birth is not a flourishing one, and it may not be worth living.
> Premise 2: Most r-strategists live lives that are filled with suffering and end shortly after birth.
> Premise 3: Most sentient individuals born into the world are r-strategists.
> Conclusion: Most sentient individuals born into the world do not live flourishing lives, and their lives may not be worth living.[7]

It's not unreasonable for one to initially react to the above argument with a sense of depressed resignation. The word's a terrible place for most sentient beings, it seems, but is there anything we can safely do about it? Isn't it too ecologically risky to try and help wild animals? In this book, I use concepts from moral and political philosophy to analyze the problem of wild animal suffering (WAS). Though many animal ethicists have the intuition that the best course of action is to leave wild animals alone,[8] I argue that a cautious, collective commitment to intervention is morally required. With enough research, it will one day be feasible for us to safely conduct large-scale, humanitarian interventions in nature. Considering how enormous the amount of WAS appears to be, I argue that we have a collective obligation to fund that research.

You may be surprised to learn that an increasingly large number of people think there are things we both can and morally ought to do about WAS.[9] Some interventions, specifically large-scale ones, will require research before they can be conducted safely and effectively, but we have strong moral reasons to fund that research. In light of the emergence of CRISPR (a new form of gene editing),[10] one especially promising line of research, in my view, is gene editing. Using a technique called 'gene drive,' we could disperse beneficial, genetically engineered traits through wild animal populations.[11]

## 1.2 Effective Altruism and Wild Animal Suffering

Some of those who endorse humanitarian intervention in nature, myself included, are professional moral philosophers with an interest in animal ethics. It's noteworthy, however, that a large number of those who endorse intervention are members of the effective altruism movement.[12] We should pause for a moment and reflect on why effective altruists are attracted to intervention in nature, as doing so will shed light on why WAS is such a pressing problem.

Effective altruism is the idea that beneficence ought to be exercised rationally.[13] It maintains that when selecting a cause to support, as well as when choosing from among the organizations devoted to that cause, one should be careful to ensure that the choice one makes is efficient. Put another way, effective altruists maintain that we should try to ensure that each additional unit of resources we devote to beneficence (whether it be money or labor) produces the most good it feasibly can. All too often well-intentioned people select a cause because it's close to their hearts, and many people fail to do the research needed to determine whether a given organization is effectively achieving its goals (or whether it's sufficiently transparent for its effectiveness to be measured, for that matter). Effective altruists maintain that the standards of effectiveness we apply to self-interested behavior should also be applied to beneficent behavior, for example, just as we would disapprovingly say that someone behaves imprudently if they buy overpriced goods when less expensive, higher quality alternatives are available, so too should we disapprovingly say that someone who limits their support to low-priority cause areas, or who donates to ineffective organizations, isn't truly being beneficent. True beneficence requires a commitment to actually doing good, not just to feeling good about one's self.

As I mentioned above, a central issue for effective altruists is cause selection. If we want our beneficent actions to produce the most good possible, which cause or causes should we pursue? To answer this question, effective altruists rely on a few main criteria, two of which are *scale* and *neglectedness*. Scale refers, more or less, to the size of the problem a particular cause is addressing. Considerations relevant to scale include the number of individuals affected by a problem, as well as the amount of suffering those individuals experience because of it.[14] Neglectedness refers to the extent to which a cause is failing to be pursued or has failed to attract interested advocates. If a worthy cause is currently being neglected relative to other worthy causes, this suggests that it's especially worth one's efforts. This is because the marginal utility of working on a cause typically decreases the more people work on it, that is, each additional unit of resources devoted to a cause will have a somewhat smaller impact than the previous unit.[15] William MacAskill uses medical doctors to illustrate this idea. He notes that, at first glance, it might seem that one can do a tremendous amount of good by becoming a medical doctor. Doctors perform a variety of important interventions, including some that save lives and thus, the average value of what a doctor does is very high. Unless one works in a community suffering from a significant shortage of doctors, though, the marginal value of one more doctor will be considerably less than the average value of a doctor. This is because, in the absence of a significant shortage, there are already enough doctors to handle the more important work that an additional doctor would end up receiving. Thus, in the event that an additional doctor is not available, someone else will perform the necessary high priority procedures. To be fair, the absence of an additional doctor has some impact. It might, for example, mean that patients with relatively minor health problems will have to tolerate longer waiting times, as there will be fewer doctors available to tend to such problems. But the fact that there are enough medical resources available to ensure that high priority health problems are addressed entails that the marginal value of a doctor lies mostly in an increased capacity for addressing low-priority problems.[16]

Considerations of scale and neglectedness have been especially key in attracting effective altruists to the WAS problem.[17] As I mentioned above, the fact that most individual wild animals born into the word are r-strategists entails that most wild animals (most sentient individuals on earth, in fact) don't live flourishing lives, and that they may not even live lives worth living (the amount of suffering they experience over the course of their lives may exceed the amount of enjoyment). As such, the scale of the problem WAS poses is absolutely massive. To be sure, there are a number of other large-scale problems, too. For example, though the number of human beings living in absolute poverty has declined considerably over the past 30 years, it remains the case that, in 2015, about 10% of the word's human population (736 million people) lacked adequate access to food, shelter, effective medicine, clean drinking water, etc.[18] Absolute poverty, that is, poverty that deprives one of the ability to consistently meet one's basic needs, thus remains a very significant problem. Similarly, the scale of the problem industrial agriculture poses for domesticated animals is enormous, and effective altruists have rightly made it one of their main focuses. According to one estimate, animal agriculture results in the slaughter of 70 billion terrestrial animals every year, most of which is attributable specifically to industrial animal agriculture,[19] and considering that the lives of these animals is characterized by confinement, sickness, malnourishment, and injury, it's fair to say that the suffering they endure before death is immense. Thus, as large as the scale of absolute poverty among humans is, the scale of the problem posed by domesticated animals in industrial agriculture is considerably larger. But even industrial agriculture is dwarfed by WAS. The word's population of wild animals is considerably larger than its population of farm animals. According to a recent estimate, the world's population of wild, terrestrial vertebrates is about one trillion, give or take a zero, and the number of wild marine vertebrates is even larger (not to mention the number of invertebrates).[20] As large as the above estimate is, though, we should keep in mind that it specifically represents the total number of (wild, terrestrial) vertebrates alive at any given moment. As a result, it doesn't include the number of vertebrates who, over a period of time, for example, 10 years, were born and then died shortly afterwards. Were we to include the many vertebrates who die shortly after being born, our population number would be far higher than one trillion. And since, as we noted, most of these animals at best fail to flourish, and at worst don't even live lives worth living, the scale of WAS far exceeds the scale of more familiar problems, even of farm animal suffering.

Neglectedness has also played a large role in attracting the attention of effective altruists. Relative to other large-scale problems, such as absolute poverty or suffering among farm animals, WAS has historically received very little attention from anyone. Of course, there is a considerable amount of interest in environmental matters, and environmental matters concern wild animals in various ways. Habitat conservation ensures that wild animals continue to have access to the resources they need to survive and procreate, and protecting biodiversity is partially about ensuring that animal species don't become extinct, as well as ensuring that the ecosystems they're a part of remain stable. Concern for wild animals' well-being, and concern for environmental values such as protecting habitats and biodiversity, are not the same thing, however. After all, protecting habitats and biodiversity is consistent with, and probably conducive to, maintaining stable r-strategist populations. And maintaining stable r-strategist populations ensures that generation after generation of doomed offspring are born into the world. Unless the protection of habitat and biodiversity are paired with separate, welfare promoting interventions, pursuing the former merely functions to protect the status quo, and as we've seen, the status quo is pretty terrible for wild animals.

The only main criterion that doesn't strongly favor selecting WAS is *tractability*. A problem is highly tractable when there are available, effective means via which to reduce it, and when it's easy to assess the extent to which our actions reduce it. A problem is less tractable when effective interventions aren't currently available but can feasibly be developed with further research.[21] Two main tractability-related worries concerning WAS are that we currently lack effective, large-scale interventions, and that the complexity of ecosystems makes it difficult predict what the effects of large-scale intervention would be. Those who are especially impressed by these worries argue that researching large-scale interventions, such as gene drives, is irresponsible. Such research is costly, and even if we succeeded in developing some interventions, anti-interventionists claim that the behavior of ecosystems is too unpredictable for us to be able to use those interventions responsibly.[22]

I'm going to address tractability-related issues in Chapter 4 and Chapter 6, so I won't discuss them at length here. I will note, however, that it isn't entirely true that we lack effective interventions at present. Human beings have had and continue to have an extensive impact on nature, enough so that human activity has arguably become the main causal factor influencing the Earth's environment.[23] Though much of our impact on nature is a harmful by-product of the continued growth of human societies, for example, climate change, some of it is deliberate and carried out for the sake of promoting environmental values, such as biodiversity or conservation, or for the sake of protecting human interests. For example, vaccination programs for wild animals have been carried out in cases where a disease affecting them threatened the health of humans or domesticated animals.[24] Though wildlife vaccination programs weren't initially developed for wild animals' benefit, it wouldn't be difficult to use them for precisely that purpose, and the fact that we've been able to safely employ them thus far suggests that we can continue to do so.

It might be tempting to think that we should restrict ourselves to familiar, reliable interventions that promise to make a relatively small (but still important) reduction in WAS. My view, however, is that limiting ourselves to such interventions fails to take WAS seriously enough. Though performing relatively small-impact interventions is something we both can and should be doing now, the sheer scale of WAS warrants researching interventions intended to have a relatively large impact on the problem. And though our fallibility and the associated risk of causing accidental ecological damage is an important consideration, I'll argue in Chapter 4 that it does not *bar* but rather functions as a *constraint* upon the justified development and implementation of high-impact interventions.

## 1.3 Chapter Outline

In Chapter 2, I critique two popular claims about nature: the claim that wild animals generally live good lives, and the claim that naturalness is a valuable property. As I described above, many wild animal species, specifically r-strategists, protect their genes by producing large numbers of uncared-for offspring, most of whom die prematurely. I argue that even on very conservative assumptions about the scope of sentience, the r-strategy entails that the vast majority of sentient animals born into the world live terrible lives. In response to the claim that naturalness is a valuable property, I argue, with some help from John Stuart Mill,[25] that naturalness is not a sufficient condition for goodness, nor a source of intrinsic value. Though it does possess extrinsic value, that value comes specifically from our epistemic limitations and thus, it fades as soon as we acquire enough knowledge to identify and successfully implement desirable interventions.

In Chapter 3, I defend the claim that WAS entails a collective duty to intervene in nature. In response to Clare Palmer's claim that because we lack a special relationship with wild animals, we also lack positive duties to them,[26] I note that only duties of care, and not duties of beneficence, are contingent upon special relationships. In response to Tom Regan's claim that we lack duties of beneficence to wild animals because wild animals are competent to manage the dangers they face,[27] I argue that most wild animal individuals are not competent, and that competence only undermines liberty-restricting interventions. I finish the chapter by arguing that pro-interventionists have good reason to focus on considerations of beneficence, rather than on distributive justice or rectificatory justice. Beneficence is a better basis than distributive justice for building democratic support for intervention, and rectificatory justice can only justify rescuing animals who won't predictably cause significant harm.

In Chapter 4, I respond to two important yet opposite criticisms of the view that we should cautiously assist wild animals on a large scale. According to one objection, even cautious intervention is unjustified because fallibility is allegedly intractable.[28] In contrast, the second objection states that we should abandon caution and intentionally destroy habitats in order to prevent wild animals from reproducing.[29] In my fourth chapter, I argue that intentional habitat destruction is wrong because negative duties are more stringent than positive duties. However, I also argue that the possible benefits of ecological damage, combined with the excusability of unintended, unforeseeable harm, suggest that fallibility should not paralyze us.

In Chapter 5, I argue that using gene drives to beneficially modify wild animal populations is a type of intervention especially worthy of research. Focusing on CRISPR in particular, I argue that the moral costs of the alternative – the perpetual interference with wild animals' liberties associated with conventional wildlife management – are far greater. Additionally, I compare a number of different goals that CRISPR could be used to try to achieve. Potential goals include making certain species, such as r-strategists or predators extinct; removing the capacity to suffer from certain animals; or changing animals' dietary and reproductive behaviors, for example, turning carnivores into herbivores or r-strategists into K-strategists. Though I argue that behavior change is ideal, I allow that removing the capacity to suffer is a promising, second-best option should behavior change prove infeasible to safely implement.

Since large-scale assistance isn't currently being provided to wild animals, fulfilling our collective obligation will first require advocating for, supporting, and/or encouraging the implementation of an assistance program. In Chapter 6, I argue that in light of Peter Singer's work on beneficence,[30] it's clear that we each have an individual duty to devote some of our resources to beneficence-related causes, WAS included. Unlike Singer, however, I argue that we're only morally required to make insignificant sacrifices. Though addressing WAS seems to require diverting advocacy-related resources away from traditional animal rights causes, I argue that it's often unnecessary to choose between them. This is because there are many courses of action that manage to simultaneously address both WAS and traditional causes. I finish Chapter 6 by arguing that illegal animal rights advocacy, though perhaps permissible in principle, is normally unjustified in practice, as legal animal rights advocacy is strategically superior.

## Notes

1 See MacArthur and Wilson (1967), and Pianka (1970). It should be noted that my use of the terms 'K-strategist' and 'r-strategist' does not mean I endorse the theory with which they are associated. The predictions that theory makes about evolved traits have often proven false, but the classificatory terminology it employs is still helpful and in common use. It should also be noted that though using the terms 'K-strategist' and 'r-strategist' in a classificatory manner suggests a dichotomy, it's more accurate to think of them as opposite points on a spectrum: some animals are clearly K-strategists, some are clearly r-strategists, but many also lie somewhere in between. For a helpful secondary source concerning the r- and K-strategies, see Jeschke et al. (2008).
2 The above paragraph is from Johannsen (2020) pp. 31–2.
3 Vitt and Caldwell (2009) p. 139, Barbault and Mou (1988).
4 For physical information about the wall lizards in their study, see Barbault and Mou (1988) p. 41.
5 Burton et al. (2002) p. 2850.
6 Vitt and Caldwell (2009) pp. 138–9. The above two paragraphs were taken from Johannsen (2017) pp. 337–8.
7 This argument outline is from Johannsen (2020) p. 32.
8 Clare Palmer calls this the 'laissez-faire intuition.' See Palmer (2010) p. 63. For an extended discussion of the laissez-faire intuition see Palmer (2010) chapters 4 and 5.
9 See, for example, Cowen (2003), McMahan (2010) and (2015), Sözmen (2013), Horta (2013), (2015), and (2017), Tomasik (2015), and Johannsen (2017).
10 Technically speaking, CRISPR just refers to a structural feature present in the genomes of different bacteria. The actual gene editing is done by an associated enzyme called Cas9 in combination with a guide RNA molecule that targets the desired part of the genome to be modified. Nevertheless, the acronym 'CRISPR' is now typically used to refer to this new form of gene editing. See Doudna and Charpentier (2014) p. 1 and p. 3. For an informative and accessible article about CRISPR see Ledford (2015).
11 The idea of dispersing a genetically engineered trait through a gene drive has been around for a while. See Burt (2003). For studies that have successfully used CRISPR to conduct gene drives see Gantz and Bier (2015), Gantz et al. (2015), and Hammond et al. (2016).
12 For an excellent book about effective altruism, see MacAskill (2015). The two major advocacy groups devoted to addressing WAS are Animal Ethics (see their website at www.animal-ethics.org/) and Wild Animal Initiative (see their website at www.wildanimalinitiative.org/).
13 For an explanation of what 'effective altruism' is, see MacAskill (2015) pp. 14–5.
14 MacAskill (2015) pp. 225–6.
15 MacAskill (2015) pp. 68–9 and 227–8.
16 MacAskill (2015) pp. 74–9.
17 See, for example, Stijn Bruers' talk "Thinking Critically about Wild Animal Suffering" at www.youtube.com/watch?v=yQF6RSSLipM
18 Information about global absolute poverty is available on the World Bank's website at www.worldbank.org/en/topic/poverty/overview
19 Schlottmann and Sebo (2019) p. 71 and p. 110.
20 Tomasik (2018).
21 MacAskill (2015) pp. 224–7.

22 See Delon and Purves (2018) pp. 244–50. For other discussions of fallibility and the problem it poses, see Singer (1975) pp. 238–9, Simmons (2009), and Ladwig (2015) pp. 297–9.
23 See, for example, Moore (2016).
24 See, for example, Buddle et al. (2011) and Rupprecht et al. (2003).
25 Mill (1885).
26 Palmer (2010) pp. 84–95 and Palmer (2015).
27 Regan (2004) p. xxxvii.
28 Delon and Purves (2018) pp. 244–50.
29 Tomasik (2016).
30 Singer (1972).

## References


Animal Ethics. Available at: www.animal-ethics.org (Accessed on July 4th, 2019).

Barbault, R., and Y.P. Mou. 1988. "Population Dynamics of the Common Wall Lizard, *Podarcis Muralis*, in Southwestern France." *Herpetologica* 44: 38–47.

Bruers, S. 2018. "Thinking Critically About Wild Animal Suffering." Available at: www.youtube.com/watch?v=yQF6RSSLipM (Accessed on July 4th, 2019).

Buddle, B.M., D.N. Wedlock, M. Denis, H.M. Vordermeier, and R.G. Hewinson. 2011. "Update on Vaccination of Cattle and Wildlife Populations Against Tuberculosis." *Veterinary Microbiology* 151: 14–22.

Burt, A. 2003. "Site Specific Selfish-Genes as Tools for the Control and Genetic Engineering of Natural Populations." *Proceedings of the Royal Society of London* 270: 921–8.

Burton, M., and R. Burton (Illustrators). 2002. "Wall lizard." In *International Wildlife Encyclopedia, 3rd Edition*. Vol. 20: 2849–50.

Cowen, T. 2003. "Policing Nature." *Environmental Ethics* 25: 169–82.

Delon, N., and D. Purves. 2018. "Wild Animal Suffering is Intractable." *Journal of Agricultural and Environmental Ethics* 31: 239–60.

Doudna, J.A., and E. Charpentier. 2014. "The New Frontier of Genome Editing with CRISPR-Cas9." *Science* 346: 1258096.

Gantz, V.M., N. Jasinskiene, O. Tatarenkova, A. Fazekas, V.M. Macias, E. Bier, and A.A. James. 2015. "Highly Efficient Cas9-Mediated Gene Drive for Population Modification of the Malaria Vector Mosquito *Anopheles Stephensi*." *PNAS* 112: E6736–43.

Gantz, V.M., and E. Bier. 2015. "The Mutagenic Chain Reaction: A Method for Converting Heterozygous to Homozygous Mutations." *Science* 348: 442–4.

Hammond, A., R. Galizi, K. Kyrou, A. Simoni, C. Siniscalchi, D. Katsanos, M. Gribble, D. Baker, and E. Marois. 2016. "A CRISPR-Cas9 Gene Drive System Targeting Female Reproduction in the Malaria Mosquito Vector *Anopheles Gambiae*." *Nature Biotechnology* 34: 78–83.

Horta, O. 2013. "*Zoopolis*, Intervention, and the State of Nature." *Law, Ethics, and Philosophy* 1: 113–25.

Horta, O. 2015. "The Problem of Evil in Nature: Evolutionary Bases of the Prevalence of Disvalue." *Relations* 3: 17–32.

Horta, O. 2017. "Animal Suffering in Nature: The Case for Intervention." *Environmental Ethics* 39: 261–79.

Jeschke, J.M., W. Gabriel, and H. Kokko. 2008. "r-Strategists/K-Strategists." *Encyclopedia of Ecology* 4: 3113–22.

Johannsen, K. 2017. "Animal Rights and the Problem of r-Strategists." *Ethical Theory and Moral Practice* 20: 333–45.

Johannsen, K. 2020. "To Assist or Not to Assist? Assessing the Potential Moral Costs of Humanitarian Intervention in Nature." *Environmental Values* 29: 29–45.

Ladwig, B. 2015. "Against Wild Animal Sovereignty: An Interest-Based Critique of *Zoopolis*." *The Journal of Political Philosophy* 23: 282–301.

Ledford, H. 2015. "CRISPR, the Disruptor." *Nature* 522: 20–4.

MacArthur, R.H., and E.O. Wilson. 1967. *The Theory of Island Biogeography* (Princeton: Princeton University Press).

MacAskill, W. 2015. *Doing Good Better* (London: Guardian Books).

McMahan, J. 2010. "The Meat Eaters." *New York Times Opinionator.*

McMahan, J. 2015. "The Moral Problem of Predation." In *Philosophy Comes to Dinner: Arguments about the Ethics of Eating*. Eds. A. Chignell, T. Cuneo, and M. Halteman (London: Routledge), pp. 268–94.

Mill, J.S. 1885. *Three Essays on Religion* (London: Longman, Green).

Moore, J. (ed.). 2016. *Anthropocene or Capitalocene? Nature, History, and the Crisis of Capitalism* (Oakland: PM Press).

Palmer, C. 2010. *Animal Ethics in Context* (New York: Columbia University Press).

Palmer, C. 2015. "Against the View That We Are Normally Required to Assist Wild Animals." *Relations* 3: 203–10.

Pianka, E.R. 1970. "On r- and K-Selection." *The American Naturalist* 104: 592–97.

Regan, T. 2004. *The Case for Animal Rights* (Berkeley: University of California Press).

Rupprecht, C.E., C.A. Hanlon, and D. Slate. 2003. "Oral Vaccination of Wildlife Against Rabies: Opportunities and Challenges in Prevention and Control." *Developments in Biologicals* 119: 173–84.

Schlottmann, C., and J. Sebo. 2019. *Food, Animals, and the Environment* (New York: Routledge).

Simmons, A. 2009. "Animals, Predators, the Right to Life, and the Duty to Save Lives." *Ethics and the Environment* 14: 15–27.

Singer, P. 1972. "Famine, Affluence, and Morality." *Philosophy & Public Affairs* 1: 229–43.

Singer, P. 1975. *Animal Liberation* (New York: Avon Books).

Sözmen, B.İ. 2013. "Harm in the Wild: Facing Non-Human Suffering in Nature." *Ethical Theory and Moral Practice* 16: 1075–88.

Tomasik, B. 2015. "The Importance of Wild Animal Suffering." *Relations* 3: 133–52.

Tomasik, B. 2016. "Habitat Loss, not Preservation, Generally Reduces Wild Animal Suffering." *Essays on Reducing Suffering*. Available at: www.reducing-suffering.org/habitat-loss-not-preservation-generally-reduces-wild-animal-suffering (Accessed on April 5th, 2019).

Tomasik, B. 2018. "How Many Wild Animals are There?" *Essays on Reducing Suffering.* Available at: www.reducing-suffering.org/how-many-wild-animals-are-there/#Explanation_of_the_estimates (Accessed on July 4th, 2019).

Vitt, L.J., and J.P. Caldwell. 2009. *Herpetology: An Introductory Biology of Amphibians and Reptiles, 3rd Edition* (Burlington: Academic Press).

Wild Animal Initiative. Available at: www.wildanimalinitiative.org (Accessed on July 4th, 2019).

The World Bank. 2019. "Poverty." Available at: www.worldbank.org/en/topic/poverty/overview (Accessed on July 4th, 2019).

# 2

# WHAT'S SO GOOD ABOUT NATURE?

## 2.1 Introduction

It is widely believed that nature is good. This belief underlies environmentalists' commitment to conservation, as well as the growing popularity of natural food and other natural consumer products. As popular as this view is, though, it suffers from a number of significant difficulties. The purpose of this chapter is to explain and critically analyze what I call 'the positive view of nature.' In light of the various problems it faces, I argue that it's largely false.

The positive view is comprised of at least two problematic claims: (a) the claim that wild animals generally live good lives, and (b) the claim that naturalness is a valuable property. In Section 2.1, I argue that claim (a) is false because only a proportionally small number of wild animals live good lives. Many wild animal species protect their genes by producing large numbers of uncared-for offspring, and as a result, the majority of (sentient) wild animal individuals die painfully and prematurely.[1] What's more, reflecting on the different sources of suffering in nature reveals that even animals who devote considerable energy to each of their offspring tend to live very difficult lives.

In Section 2.3, I argue that only a qualified version of claim (b) is plausible. Drawing on John Stuart Mill,[2] I argue that naturalness is not a sufficient condition for goodness, nor a source of intrinsic value. Though I concede that it possesses extrinsic value, its value comes specifically from our epistemic limitations,[3] and this kind of value fades as soon as we acquire enough knowledge to identify and successfully implement desirable interventions.

In Section 2.4, I seek to explain why the positive view of nature has initial appeal. I suggest that its initial appeal is, to some extent, traceable to our tendency to use unrepresentative examples when imagining life in the wild,[4] and, to some extent, traceable to the contingent aesthetic appeal that natural things sometimes have.

## 2.2 Do Wild Animals Generally Live Good Lives?

### *2.2.1 The r-Strategy*

It's commonly believed that nature is idyllic. In other words, many believe that nature is a pleasant place where animals typically live flourishing, 'wild' lives.[5] To be fair, I don't think any reasonable person would deny that nature contains some amount of violence and suffering. The predator–prey relationship is pervasive, after all, and it's obvious enough that being eaten by a predator is a terrible way to die. Still, enough people believe that, on average, wild animals live flourishing lives. If this belief were true, then there would be one very significant sense in which nature is good – it would be good in that it contains a high level of net well-being.

As noted in Chapter 1, however, suffering is pervasive throughout nature, and a major reason for that is the prevalence of the r-strategy. Many lizards, amphibians, fish, and small mammals, protect their genes by producing large numbers of offspring, rather than by producing small numbers of offspring and investing a large amount of energy into each infant.[6] For any group of animals whose population is stable, protecting one's genes via the r-strategy entails that the vast majority of one's offspring will die at a very young age since, on average, only one offspring per parent will survive to maturity. And the death of one's offspring will usually occur as a result of causes that, for any sentient being, are very painful, for example, predation, disease, starvation, exposure, etc. In light of these facts, we noted that the below argument seems sound;

> Premise 1: A life that's filled with suffering and ends shortly after birth is not a flourishing one, and it may not be worth living.
> Premise 2: Most r-strategists live lives that are filled with suffering and end shortly after birth.
> Premise 3: Most sentient individuals born into the world are r-strategists.
> Conclusion: Most sentient individuals born into the world do not live flourishing lives, and their lives may not be worth living.

There are, of course, a number of empirical questions concerning the extent of r-strategist suffering, for example, the question of which r-strategists are sentient and thus, are capable of feeling pain at all. Insects, for example, may very well not be sentient, and it's possible that some r-strategists don't become sentient until a later stage in their lives, for example, amphibian r-strategists who must initially go through a larval stage. Additionally, some philosophers and cognitive scientists draw a distinction between 'mere pain': a purely sensory experience, and 'suffering': the state one's in when one minds or is disturbed by their pain,[7] and wonder what must be true of an organism for it to be capable of suffering. Though the available evidence suggests that non-human mammals can suffer, it's somewhat uncertain whether other non-human animals generally can.[8] The possibility that many r-strategists can only feel 'mere pain' – a cognitive state that's less unpleasant, and which presumably has less moral significance, than suffering – suggests that the lives of r-strategists may not actually be as bad as they seem.

The empirical uncertainty surrounding the extent of r-strategist suffering is important to note, but it should also be noted that the r-strategy has significant implications for wild animal suffering (WAS) on even conservative assumptions about the scope of sentience and the scope of the capacity to suffer. Suppose, for the sake of argument, that we were to restrict the scope of our moral attention to mammals, since they can certainly suffer and since suffering is certainly a morally significant mental state.[9] Even with this restriction, it remains the case that the vast majority of individuals born into the moral community are r-strategists. After all, the majority of mammals are small mammals, and small mammals tend to be closer to the r side of the r/K spectrum than large mammals.[10] Consider, for example, the meadow vole (*Microtus pennsylvanicus*). According to one study, 88% of meadow vole young die during their first month of life, and as a result, the average meadow vole only lives for 0.7 months.[11] Furthermore, the cause of death for meadow vole infants is, in the majority of cases, a rather painful one: specifically, predation.[12] Though not all small mammals are r-strategists, it remains true that the r-strategy is common enough among mammalian species. And so long as the r-strategy is common among mammals, then the above argument remains sound even when we replace the term 'sentient individuals' with 'mammalian individuals.' In other words, the r-strategy implies that the majority of mammalian individuals born into the world do not live flourishing lives, and that their lives may not be worth living.[13]

The r-strategy is not, of course, the only source of suffering in nature. This seems obvious enough when we note that other causes, such as predation, are directly responsible for the death of r-strategist infants, that is, that the r-strategy is an indirect cause of suffering and death, rather than a proximate cause. Furthermore, r-strategist infants aren't the only wild animals who experience a low level of welfare. Most (sentient) K-strategist animals and r-strategist adults endure a considerable amount of suffering from a variety of sources, and in the next two subsections, I identify many of those sources and explain the relationships between them.[14]

### 2.2.2 Predation, Weather, and Food Scarcity

Weather and food scarcity are two major, interrelated factors affecting the amount of suffering and death in nature. The availability of food plays a large role in limiting population sizes, and the weather, in turn, plays a large role in affecting the availability of food. When food is scarce, it limits population size mostly by increasing mortality among a species' young, that is, children born into not only r-strategist but also K-strategist species, will starve to death.[15]

Weather is also an independent source of suffering for wild animals, as fluctuating temperatures and weather conditions are a direct cause of physical discomfort. In cases of extreme heat or extreme cold, some animals die of exposure.[16] But even normal, non-lethal fluctuations can cause considerable discomfort. Consider, for example, the fish species *Lota lota*. This species is a member of the Gadidae family, a group of marine fish. Unlike the other members of this family, though, *Lota lota* mainly inhabits cold rivers and streams in North America and Northern Europe and thus, it must endure far higher temperatures in the summer than other Gadidae. *Lota lota* manages to survive during periods of heat by reducing both its physical activity and its metabolic rate, but the fact that it manages to survive during these periods doesn't mean that it's happy during them. Quite the opposite: the summer heat is a source of considerable stress for these fish.[17]

In addition to weather, predation interacts with food scarcity. When food is abundant, members of prey populations have the luxury of being able to reside in areas where the risk of predation is lower, for example, areas that afford them a lot of cover, but when food is scarce, it becomes necessary to forage more and to thus expose one's self to greater risk. As a result, members of prey populations often find themselves in a damned if you do, damned if you don't situation: they have to choose between possible starvation, on the one hand, and exposing themselves to attack, on the other.[18]

A final interacting factor I'll mention in this subsection is infectious disease. Wild animals are vulnerable to various infectious diseases, many of which are painful and debilitating. As such, disease is, to some extent, an independent source of suffering for wild animals. Additionally, though, it has a relationship with food scarcity. Animals in poor physical condition, for example, animals who are starving because of food scarcity, are more vulnerable to pathogens. And contracting a pathogen, in turn, drains the animal further by making their immune system work harder, or by making it physically more difficult for the animal to forage. The result is a vicious circle in which poor physical condition leads to disease, which leads to a worsening of physical condition, which in turn leads to even further vulnerability to disease.[19]

### 2.2.3 Predation, Stress, Injury, and Parasites

In the above subsection, it was noted, among other things, that predation interacts with food scarcity. It's noteworthy that predation is associated with a number of other sources, too. Of all the sources predation interacts with, stress is perhaps the most constant. As distressing and painful as it is to be killed by a predator, the stress associated with being perpetually at risk of attack is perhaps nearly as bad. Predator-induced stress often impacts prey species' reproduction, but it can affect them in a variety of other ways as well.[20]

An additional factor that interacts with predation is injury. Predators must, of course, injure their prey in order to kill them, but often enough the injuries sustained during a predator attack prove to be non-lethal, and the targeted prey manages to either fend off or escape from their attacker. For the animals lucky enough to survive an attack, the injuries suffered may include anything from skin abrasions to broken bones and punctured flesh.[21]

Wild animals sustain injuries from more than just predation. One particularly nasty source of physical trauma is forced copulation, that is, rape. Rape and other kinds of sexual violence have been observed in a variety of animal species, including different primates and waterfowl.[22] According to a study of wild orangutans, most cases of intercourse between females and subadult males involve the use of coercive violence by the latter, and nearly half of all cases between females and adult males involve coercive violence, too.[23]

Injuries, whether caused by a predator attack or by some other cause, often become infected in the wild. Bacterial infections are common, of course, but parasitic infections can also occur. For example, wild animals' wounds sometimes become infected with Myaisis: the infestation of tissue by fly larvae.[24] A particularly nasty version of Myaisis was, until recently, quite common among North American wild and farmed mammals. The New World screw fly would lay its eggs on some wounded animal, and upon hatching, the fly's larval offspring would proceed to consume the flesh of their host. Though the larvae eventually left their host after eating their fill, they made sure to emit a pheromone before doing so: one that would attract new female flies to the growing wound. As a result, animals infected with screw fly larvae would slowly but surely be eaten alive. Fortunately, thanks to human intervention, the New World screw fly was virtually eradicated from North America at the beginning of the millennium. The technique used to eradicate it proved less effective in South America, though, where New World screw flies still exist, and other species of screw fly also continue to exist elsewhere in the world.[25]

Myaisis, especially the sort caused by screw flies, is a particularly gruesome sort of parasitic infection, but parasitic infections of many kinds are common among wild animals. Examples include *Trichinella*, *Echinococcus spp.*, *Leishmania*, *Sarcoptes scabiei canis*, *Babesia*, *Trichomonosis*, and *Haemosporida*.[26]

### 2.2.4 Are Wild Animals' Lives Worth Living?

In the above sections, note was taken of various sources of suffering in nature. Some of them affect only some animals some of the time, but many of them are pervasive. As bad as WAS is, though, it's likely that many wild animals have lives worth living. After all, nature doesn't only contain suffering, it contains various sources of pleasure, too. For social animals, these pleasures include things like playing games, grooming one another, caring for a family, etc. And for any sentient animals, the pleasure of finding and consuming food, or of successfully exercising one's talents and abilities, will be significant.[27] For wild animals that live long enough to enjoy an array of pleasures, then, it may very well be the case that the pleasant experiences they enjoy exceed the negative ones. Still, we should keep in mind that a life worth living is not necessarily a flourishing one.[28] Life can still be quite difficult without being so bad as to get to the point where the amount of suffering one experiences exceeds the amount of enjoyment. For example, the conditions faced by people in absolute poverty are very difficult. Life is hard when one not only has less than others, but is perpetually malnourished, lacks easy access to clean water, has inadequate shelter, has inadequate access to medical assistance, etc. Still, it's likely that most of those in absolute poverty live lives worth living. Most still have the opportunity to form friendships, have families, learn skills, and develop talents, etc.[29] Given that they face similar hardships, it's perhaps appropriate to think that the lives lived by wild animals who survive to maturity are, on average, similar in quality to (but probably still somewhat worse than) the lives lived by people in absolute poverty.

As we already know, though, the vast majority of wild animals die during infancy and thus, do not get to enjoy an array of rewarding experiences. Some of them manage to consume a bit of food, but to a large extent they just crawl around uncomfortably and then die painfully after a short while. I think it's plausible (although not completely certain) that their lives aren't worth living.

A large part of what's bad about the r-strategy is the suffering it causes. Additionally, though, it may be the case that premature death is independently bad for r-strategists. In the human case, we think that premature death is bad for the person who dies regardless of whether it involves suffering, and extending that judgment to wild animals entails that the death of r-strategist infants is bad for them independently of the extent to which their death is painful. Should we extend this judgment to wild animals, though? It could be argued that a painless death is only bad for a being if that being has future-oriented desires.[30] The aspiration to have a rewarding career, to have a family, to write a book, etc., are all long-term goals that concern a person's future, and dying would frustrate their fulfillment. If a being lacks future-oriented desires, though, then arguably a painless death cannot be bad for them, and it's unlikely that r-strategist infants are cognitively complicated enough to have more than very immediate desires, for example, the desire to eat when hungry, sleep when tired, etc. Of course, human infants lack future-oriented desires, too, as do some human beings with very severe cognitive disabilities. The future-oriented desires view thus unpalatably implies that painlessly killing such humans is not bad for them.

A better view is Don Marquis's, that is, the view that death is bad for a being when it deprives it of a valuable future.[31] In this view, it isn't necessary for death to frustrate a victim's present desires. It's enough that death deprives its victim of the valuable experiences it would have had were it to not have died. This account includes r-strategists and has the virtue of not excluding human infants or people with very severe cognitive disabilities. However, whether it entails that premature death normally harms r-strategist infants depends on whether those few r-strategists who reach maturity normally have lives worth living. I've suggested that, on average, wild animals who reach maturity probably do have lives worth living, even though their lives are very difficult. If I'm right, then on Marquis's account, the premature death of r-strategist infants is bad for them independently of the suffering they experience when dying.

One might object that Marquis' account also entails that aborting fetuses, even early-term ones, is wrong, and that his account should be rejected for that reason. Indeed, Marquis himself thought that his account implied that abortion is typically wrong. For my part, though, I don't think that Marquis is right about the implications of his account. As he notes himself, applying his account requires that we be able to identify a subject of harm,[32] and I don't think that it's correct to think of non-sentient, early-term fetuses as potential subjects of harm. To be a subject of harm, one must be – or in the case of comatose patients, have been – sentient. It isn't enough that something will become sentient at some point in the future. To see if you agree, imagine a robot that's been programmed to become sentient after a certain period of time has passed (a day let's say). Suppose also that on the back of this robot is a 'cancel' button that, if pressed, will delete the sentience-related part of its programming. Though it would be wrong to press this button after the robot becomes sentient, would it be wrong for the robot's creator to prevent it from becoming sentient by pressing the button beforehand? Presumably not, and the reason is because the robot can't be harmed until it's a sentient being. If sentience is necessary for this sort of subject-hood, then we can consistently embrace Marquis's account of the (suffering independent) wrongness of death without accepting that it's wrong to abort early-term fetuses.

If r-strategists generally don't live lives worth living, and if the premature death of r-strategist infants is also bad for them independent of the suffering it causes, then the r-strategy causes an astronomical amount of harm in the world. But even if it turns out that premature death is not independently harmful, and even if it turns out that most r-strategists technically live lives worth living (that their enjoyment exceeds their suffering), it would remain true that most wild animals born into the world live terrible lives. It isn't okay that most wild animals live terrible lives – not if we're able to do something about it.

## 2.3 Is Naturalness Valuable?

The r-strategy entails that the first part of the positive view of nature is false, that is, that nature is not idyllic. The second claim comprising the positive view of nature is that naturalness is a valuable property, and that we therefore have reason to protect, promote, and do that which is natural. As John Stuart Mill notes,

> That any mode of thinking, feeling, or acting, is "according to nature" is usually accepted as a strong argument for its goodness. If it can be said with

> any plausibility that "nature enjoins" anything, the propriety of obeying the injunction is by most people considered to be made out: and conversely, the imputation of being contrary to nature, is thought to bar the door against any pretention on the part of the things so designated, to be tolerated or excused; and the word unnatural has not ceased to be one of the most vituperative epithets in the language.[33]

Though Mill's essay "Nature" was written in the 1800s, it's not hard to find contemporary cases where naturalness is treated as a value. For example, the thought that naturalness is valuable underlies environmentalists' commitment to wilderness conservation, as well as the preference a growing number of consumers have for natural products, for example, natural food, supplements, cosmetics, etc.

For his part, Mill is very critical of the idea that naturalness is valuable. More specifically, Mill distinguishes between two common senses of the term 'nature,' and he argues that regardless of which sense one has in mind, it's absurd to think that 'nature' is good. The first sense he distinguishes is the sense that chemists and physicists (and maybe biologists) have in mind when they speak of the natural word. In this sense, 'nature' includes literally all phenomena.[34] It's this sense of the word that people have in mind when they occasionally point out that human beings are a part of nature, too, and the problem with it is that it's too expansive to be a value. To be meaningful, the directive to protect, promote, or do that which is natural, must refer to some subset of the things it's possible to protect, promote, or do. But in the first sense, to protect anything is to protect something natural, and to promote or do anything is to promote or do something natural. Because it includes everything, the wide sense of 'natural' cannot be used as a criterion to distinguish what we should protect from what we shouldn't protect, what we should promote from what we shouldn't promote, or what we should do from what we shouldn't do.[35]

The second sense of 'nature' that Mill distinguishes is much narrower in scope. It specifically refers to that which exists or occurs independent of human agency.[36] This second sense is a more likely candidate for what people have in mind when they invoke naturalness as a value. Returning to our previous examples, consumers who prefer to buy 'natural' products are specifically interested in products that, though partially produced by human agency, are further removed from it, for example, products comprised of non-synthetic or non-artificial ingredients. Similarly, environmentalists interested in conservation are specifically interested in protecting areas that aren't the product of human activity from human influence, for example, protecting swamps, forests, etc., from human encroachment. There are a number of philosophical questions about how, precisely, the second sense of nature should be understood, as well as about what protecting naturalness involves. For example, we might wonder whether protecting naturalness always requires us to refrain from intervention, or whether intervening in order to counteract the effects of human activity is compatible with it, for example, intervening in order to remove an invasive, human-introduced species; or to protect an area from the ecological effects of climate change.[37] Complications such as this lie beyond the scope of the present discussion, though.

According to Mill, the problem with the second sense of 'nature' is that most of what we do presupposes that much of what's natural is bad. After all, if we bother to try and accomplish something, then we presumably think that what we're trying to do ought to be accomplished. But if the absence of human agency is valuable, then presumably we shouldn't be trying to accomplish things. For example, it's perfectly natural to be exposed to the elements and to experience the discomforts associated with them. Feeling cold when the weather's cold, or being wet when the weather's wet, are natural states. But since being in them causes us discomfort, they aren't good states. Similarly, it's perfectly natural for people to die painfully from disease and infection, but since suffering and death are both bad, preventing them is good regardless of whether their causes are natural.[38] Of course, some human projects are bad, and some deviations from nature are bad, too, for example, dropping a nuclear bomb on a tract of wilderness. Mill's point is just that the mere fact that something is natural does not mean that it's good.

Another area where deviating from nature is often good is with respect to what some call 'instincts' or 'natural impulses.'[39] This subject is a difficult one, in part because it's often controversial whether a particular impulse is biological or sociological in origin, but also because there's controversy concerning the conditions which must be met to autonomously act upon a biological impulse.[40] Notwithstanding these issues, though, there are a number of impulses we can point to that are likely biological in origin, and it's common sensical enough to note that some tension exists between exercising agency and being unable to resist an unwanted impulse. As such, exerting control over which impulses motivate one's behavior, that is, resisting impulses one disapproves of and behaving in accordance with impulses one approves of, expresses one's agency.[41] But if it's good to do what's natural, then we presumably ought not to resist any of our natural impulses. Motives such as fear, or the desire to dominate others, would have to be embraced on a pro-natural impulse view.[42]

It might be objected that the above set of Millian arguments merely show that naturalness isn't a sufficient condition for goodness. Since values other than naturalness presumably exist and can come into conflict with it, natural things can sometimes be bad, all things considered, even if their naturalness is a source of value. Effectively critiquing the claim that naturalness is a valuable property requires showing that, all things being equal, naturalness does not increase something's value.

To test whether naturalness is a source of value, let's compare cases where the presence or absence of naturalness is the only relevant difference. In our first case, suppose that someone is bitten by a mosquito and dies from a mosquito-borne illness – Malaria, let's say. In our second case, suppose that someone dies from the same illness, but that they didn't contract it from a random mosquito.

Instead the illness was inflicted upon them by someone who wished to cause harm, for example, via a syringe full of malarial blood. Does the fact that Malaria was contracted naturally in the first case make it less bad than in the second case? I think so, actually. The fact that Malaria was intentionally inflicted in the second case means that there's someone to blame for it, and harms caused by blameworthy agents do seem to be worse than otherwise similar natural harms. However, I don't think this observation shows that naturalness is a source of value. Instead, it shows that agency amplifies an effect's significance, that is, that bad effects are worse when they're the product of agency, but also that good effects are better when they're the product of agency. To see this, let's compare a second pair of cases. In the first case, someone is accosted by a thug but manages to escape without harm because the ground is wet with rain and the thug accidentally slips and falls. In the second case, someone is accosted by a thug but manages to escape without harm because a brave bystander intervenes and manages to deter the thug. If naturalness were a source of value, we'd have to say that the first case of escaping without harm is better than the second case, since the first case is largely the product of natural happenstance (the fact that it rained recently). If anything, though, the second case of escape seems better than the first, since it was produced by the heroism of a praiseworthy rescuer.

In the above paragraphs, I hope to have established (with some help from Mill) that naturalness is neither sufficient for goodness, nor an intrinsic source of value. A remaining possibility, however, is that naturalness is, at least sometimes, an indirect source of value. In fact, some philosophers have maintained that natural spaces are valuable largely because they're needed to satisfy the interests of wild animals, that is, that they're valuable because they're habitats.[43] I'm going to discuss the relationship between WAS and habitat in much greater detail in Chapter 4, but I'm bringing it up here because it suggests some convergence between the view that naturalness is a valuable property, and the view that nature is idyllic. In particular, if it were true that wild animals are living reasonably good lives, and if it were also true that human intervention threatens those lives in and of itself, then we'd have good indirect reasons to refrain from impinging upon the naturalness of wild animal habitats.

As we've already seen, however, the view that nature is idyllic is false. As a result of the r-strategy, the vast majority of wild animals fail to flourish, and their lives may not even be worth living. More generally, and as Richard Dawkins observes, evolution doesn't function to promote genes that produce happiness-conducive traits. What it promotes are genes that propagate themselves, that is, genes which produce traits that ensure an organism is reproductively successful.[44] The r-strategy is a clear example of a trait (or set of traits) that protects a parent's genes specifically at the expense of its offspring's well-being, but there are many other instances of traits that protect genes at the expense of well-being. For example, male Peacocks are burdened with very heavy, movement encumbering feathers, and male songbirds spend dangerous amounts of time singing (dangerous because singing for long periods of time attracts predators and exerts a lot of energy). The reason male Peacocks and songbirds are like this is because female members of their species find it attractive, which makes these traits conducive to reproductive success.[45]

Though wild animals do not generally live good lives, there may be another, more plausible sense in which nature is indirectly valuable. It may be the case that nature is valuable because of our fallibility.[46] After all, nature is very complex, and as a result, it's often difficult to accurately predict the effects of human intervention. In light of our epistemic limitations, perhaps we ought to be hesitant to intervene, as it's hard to know what harms we might inadvertently cause to sentient individuals.[47] For example, it might be said that the problem with highly artificial foods is that we don't know exactly how eating them affects us in the long term. Maybe they contribute to the likelihood of developing cancer? And it might be said that the problem with large-scale interventions in the wilderness is that we can't predict exactly how they'll affect the animals living there. Though natural processes such as predation and the r-strategy are certainly bad for animals, interfering with those processes by, for example, separating predators from prey or by controlling r-strategist reproduction, risks creating an even worse situation.[48] If nature is valuable because intervention risks hurting either ourselves or animals, then it's valuable because of its connection with the well-being of sentient creatures. Furthermore, this sort of value is not only indirect but also counterfactual. For example, though it would be false to claim that wild animals generally live good lives, we might nonetheless note that interference risks making their situation worse, and that wild animal habitats are thus better as they are than as they would be in a counterfactual scenario where human intervention has led to disaster.

For my part, I think fallibility is a significant consideration. However, I don't think that it should be paralyzing. Fallibility is a constraint on justified intervention, rather than a bar to it. The reason is simply because the indirect value associated with fallibility disappears when we acquire the knowledge necessary to manage the risks of interference. For example, many conditions that we normally use medicine to treat once carried with them the sort of counterfactual value that's associated with fallibility. Obviously, poor health is bad, and medical intervention is meant to improve health, but when people knew little about how the human body works or about how to treat illness and injury, interventions such as administering a drug or performing surgery often carried a high risk of worsening the patient's condition. Once medicine became more advanced, though, the range of situations where it became appropriate to intervene significantly increased. With respect to interference with natural processes, then (or the creation of artificial food, for that matter), the appropriateness of intervention is relative to the amount of research we've done and our corresponding state of knowledge. The more we know about the likely effects of different forms of intervention, the easier it becomes to identify desirable interventions. It's worth noting that in Chapter 4, I defend the view that fallibility is merely a *constraint* on justified intervention at greater length.

## 2.4 Why is the Positive View of Nature so Popular?

In the present chapter, I argued that the positive view of nature is largely false. Nature is clearly not idyllic, and naturalness is only valuable in so far as our fallibility makes interference unwise. In light of the view's popularity, though, it's worth wondering what people see in it. Why is the positive view of nature so appealing for so many people?

To some extent, the positive view of nature is explained by survivorship bias.[49] In particular, the belief that nature is idyllic reflects a failure on most people's part to use representative examples when imagining life in the wild. By and large, people tend to think of large, K-strategist mammals when they imagine examples of wild animals, for example, deer, bears, elephants, giraffes, etc. Large K-strategists are noticeable and they share numerous similarities with us, for example, they're sociable, have families, etc., so it's 'natural' enough that we're predisposed to think of them. But since most animals born into the world are r-strategists, our impression of what life is like for K-strategist mammals is unrepresentative of what life is like for wild animals in general. In fact, it's probably not even representative of what life is like for wild K-strategists themselves, as it isn't always obvious when wild animals feel ill or are in pain. One reason for this is that many prey species hide signs of sickness or injury in order to avoid giving predators the impression that they're vulnerable.[50] To be fair, r-strategists sometimes come to mind, too, for example, bull frogs, crocodiles, etc. But when we think of bull frogs or crocodiles, most of us specifically envision them as mature adults, rather than as infants. As we already know, mature r-strategists are just as unrepresentative as K-strategists since the vast majority of r-strategists die during infancy.

Another consideration that helps to explain the positive view's appeal is that many natural things are aesthetically pleasing, for example, natural landscapes, wild flora, etc. It should be noted, however, that the association between naturalness and aesthetic value is contingent. It's not hard to think of natural things that aren't aesthetically pleasing, for example, disease, starvation, parasitism, etc. What's more, it's not hard to think of unnatural things that are aesthetically pleasing, for example, paintings, sculptures, etc. That there is an abundance of natural but not aesthetically pleasing things, and unnatural but aesthetically pleasing things, suggests that when natural things are aesthetically pleasing, the reason isn't because they're natural, but rather because they possess some other feature or set of features, for example, vastness, complexity, bright colors, etc.

One might object that naturalness supplies *some* aesthetic value, even if it is neither necessary nor sufficient for something to be aesthetically pleasing overall. In support of this claim, one might argue that intrinsic aesthetic value is needed to explain why restoration cannot fully replenish a formerly natural space's value.

Robert Elliot argues for a conclusion along these lines in his paper "Faking Nature."[51] According to him, restoring a forest, swamp, etc., is analogous to creating a forgery. Just as a forged piece of art lacks the full value of the original, a restored environment lacks the full value of the natural space it replaces. Elliot claims that the loss of value is attributable to the fact that both the forgery and the restored environment have the wrong history.[52] As convincing and as beautiful as some forgeries are, part of an artwork's aesthetic value is traceable to the fact that it was created by a particular artist who possessed a particular set of intentions, and who was situated in a particular political, social, and historical setting. And though a restored forest might be both beautiful and similar to the original forest, part of the original's value (according to Elliot) is traceable to the fact that it was created by natural processes. Since both the forgery and the restored forest were created in ways that lack the relevant, value-conferring features, their value is less than the value of the originals.

I share Elliot's intuition that a restored environment is, in some respect, worse than the original. However, I don't think this intuition suggests that naturalness possesses intrinsic value. For one, the intuition Elliot draws on concerns the value of an imitation relative to what it imitates. Though it strikes me as plausible to claim that imitations are, in one respect, inferior to that which they imitate, this claim only suggests that *originality* is a source of value, not that *naturalness* is a source of value. Furthermore, Elliot's intuition concerning restoration, in particular, is about the value of a replacement relative to what it replaces. Though it strikes me as plausible to claim that particular, valuable things are sometimes irreplaceable, for example, that a person one loves cannot simply be substituted with someone else; or that a beautiful statue cannot fully be replaced by another, equally beautiful statue; this claim only suggests that the value of a valuable particular is not always or entirely reducible to the valuable properties it possesses.[53] In other words, valuable particulars, at least sometimes, matter in and of themselves, and not just because they possess value-conferring properties such as beauty, intelligence, etc. To test whether naturalness is a source of intrinsic, aesthetic value, it's necessary to design a case that isolates naturalness from evaluatively relevant variables such as whether something is an imitation or whether it's a replacement. A case where something natural possesses more or less value than something unnatural is only a source of philosophical evidence if the items being compared are otherwise analogous. For example, consider two islands, both of which are initially devoid of life but which come to be populated by a variety of aesthetically pleasing grasses, shrubs, and flowers. On the first island, flora began to grow because a passing storm blew various seeds onto it. On the second island, flora began to grow because certain people decided to go there and plant them. If naturalness is a source of value, then the first island should be more valuable than the second, since the process that brought flora to the island was natural. However, it isn't clear that the first island is better than the second. If anything, the second island seems better, as its beauty was produced by the efforts of praiseworthy planters.

Though naturalness if not a source of intrinsic value, it of course remains true that many natural things are aesthetically pleasing. What's more, aesthetic value is worth taking note of wherever it exists, in part because it can create indirect moral obligations. We might, for example, have an obligation to avoid destroying beautiful things, such as antique paintings, because doing so would deprive others of the opportunity to enjoy them. However, the obligation to avoid destroying beautiful things is not a particularly strong one, comparatively speaking. If one were ever forced to choose between assaulting a person and assaulting a painting, one should obviously assault the painting.

## 2.5 Conclusion

In conclusion, this chapter has critiqued what I call 'the positive view of nature,' that is, the view that (a) wild animals generally live good lives, and (b) that naturalness is a valuable property. In Section 2.2, I argued that claim (a) is false because only a proportionally small number of wild animals live good lives, and in Section 2.3, I argued that only a qualified version of claim (b) is defensible. Though naturalness is counterfactually valuable, it is only so because of our epistemic limitations. Finally, in Section 2.4, I suggested that the positive view's initial appeal is traceable to the influence that survivorship bias has on us when we imagine life in the wild, and to the contingent aesthetic appeal that some natural things have.

The moral upshot of this chapter is that we shouldn't condemn actions that impinge upon nature, but that we should also perform them cautiously. Much has been and is to be gained from improving formerly natural states of affairs, but we must be careful not to cause excessive harm in the process.

## Notes

1 Animals who protect their genes by producing large numbers of uncared-for offspring are normally referred to as r-strategists. See MacArthur and Wilson (1967), and Pianka (1970). For a helpful secondary source, see Jeschke et al. (2008).
2 Mill (1885).
3 Schlottmann and Sebo (2019) p. 15.
4 Bruers (2018) and Horta (2010) section 4.
5 For a discussion and critique of this view, see Horta (2010).
6 See MacArthur et al. (1967), Pianka (1970), and Jeschke et al. (2008).
7 See, for example, Farah (2008). Upon reflection, I'm not sure that it's plausible to draw a sharp distinction between 'mere pain' and 'suffering.' Unlike nociception in non-sentient organisms, mere pain is supposed to be a conscious experience, rather than a purely reflexive reaction. In light of this, it doesn't seem possible for mere pain to motivate avoidance behavior of a sort that's distinguishable from what nociception causes, unless those who experience it also mind or are somewhat disturbed by it. It may be more plausible to think of the distinction in terms of a threshold. In this view, suffering is something one experiences when one cares sufficiently about or

is sufficiently disturbed by their pain, whereas one experiences mere pain when the extent to which one cares about it falls below the threshold.

8 See Shriver (2006) and Farah (2008).

9 Of course, mere pain is also a morally significant mental state, as is pleasure, so it isn't very plausible to restrict the scope of moral attention to mammals. I entertain this restriction only for the sake of argument.

10 See the discussion of small mammals' life histories in chapter 1 of Stoddart (1979).

11 Getz (1960) p. 398.

12 Getz (1960) p. 397.

13 The above argument outline, and the two paragraphs following it, were largely taken from Johannsen (2020) pp. 32–3.

14 The bibliographies available from Animal Ethics at www.animal-ethics.org were invaluable for completing the below subsections.

15 See White (2008).

16 See Sládek (1881).

17 See Hardewig et al. (2004).

18 See Anholt et al. (1995), McNamara and Houston (1987), and Sinclair and Arcese (1995).

19 Beldomenico et al. (2008).

20 See, for example, Boonstra R. (1998), Creel et al. (2007), and Dwyer (2004).

21 For some examples, see Heithaus at al. (2002) and Reimchen (1988).

22 Clutton-Brock and Parker (1995), McKinney and Evarts (1998).

23 See Clutton-Brock and Parker (1995) p. 1346. For the study cited by Clutton-Brock and Parker, see Mitani (1985).

24 See Francesconi and Lupi (2012).

25 See Kevin Esvelt's discussion of screw fly larvae in Esvelt (2018). See also Vargas-Terán et al. (2005).

26 For a list of parasites that afflict wild animals, and for various references about those parasites, see Animal Ethics (2019[BIB-003]).

27 See Maris and Huchard (2018) pp. 124–8.

28 In my earlier work, it was hasty of me to grant that K-strategists generally flourish. See Johannsen (2017) p. 338.

29 My description of absolute poverty is a slightly modified version of the one I provided in Johannsen (2020) at pp. 34–5.

30 For a discussion of future-oriented desires and their relationship with the badness of death, see Singer (1993) p. 95 and p.125.

31 See Marquis (1989).

32 Marquis (1989) pp. 201–2.

33 Mill (1885) p. 11.

34 Mill (1885) pp. 4–6.

35 Mill (1885) pp. 15–9.

36 Mill (1885) pp. 7–8. It's worth noting that there are a number of questions concerning how, exactly, the narrow sense of 'nature' should be understood. For example, we might ask how important it is for what's caused by human behavior to be intentional. Is a landscape that's intentionally altered by human behavior less natural than a landscape that's unintentionally altered by it? Or is the extent of the alteration, i.e., the scale of the impact, the more significant factor? For a discussion of such issues, see Hettinger (2018) pp. 68–9.

37 For a discussion of interventions that may increase naturalness, see Hettinger (2018) p. 69.
38 Mill (1885) pp. 19–21 and pp. 28–31.
39 Mill (1885) pp. 43–59.
40 Though Hume thought that free will is simply a matter of not being prevented from doing what one desires to do, many philosophers argue that one's desires must meet certain conditions in order for the actions that flow from them to be free. For Hume's thoughts on freedom of will, see Hume (1993) pp. 53–69. For alternative views, see, for example, Frankfurt (1971) pp. 14–20, and Dworkin (1976) pp. 25–6.
41 It's worth noting that philosophers disagree about whether control is necessary for agency. According to Frankfurt's classic account (see endnote 40), what matters is whether the agent approves of the desires that motivate its behavior, not whether it could have acted otherwise.
42 See endnote 39.
43 Indeed, some philosophers have argued that wild animals' interest in habitat entails that they have rights to habitat that are analogous to property rights. See Hadley (2015) and Milburn (2017).
44 Dawkins (1995) pp. 105–6.
45 Dawkins (1995) p. 120.
46 Schlottmann and Sebo (2019) p. 15.
47 For an extensive discussion of how this problem bears on the issue of WAS, see Delon and Purves (2018).
48 It's worth noting that any unintended effects caused by large-scale interventions wouldn't necessarily be net-negative. In fact, they could potentially be net-positive. I discuss the implications of this possibility in Chapter 4.
49 See Bruers (2018) and Horta (2010) section 4.
50 That prey animals hide signs of sickness or injury is sometimes discussed in veterinary research. See, for example, Dykes and Orr (2006), and Tizard (2008).
51 Elliot (1982). To be fair, I'm oversimplifying Elliot's position a bit. He considers responses to nature to be different from aesthetic responses because he thinks that aesthetic objects are necessarily intentional artifacts (see p. 90 in his article). However, his analogy between restoration and forgery suggests that he thinks responses to nature are at least quite similar to aesthetic responses.
52 Elliot (1982) pp. 83–9.
53 For a discussion of the idea that the value of particulars is not reducible to the valuable properties they possess, see Cohen (2013).

## References

Anholt, B.R., and E.E. Werner. 1995. "Interaction Between Food Availability and Predation Mortality Mediated by Adaptive Behavior." *Ecology* 76: 2230–4.

Animal Ethics. Available at: www.animal-ethics.org (Accessed on July 7th, 2019).

Animal Ethics. 2019. "Antagonism in Nature: Interspecific Conflict." Available at www.animal-ethics.org/wild-animal-suffering-section/situation-of-animals-wild/interspecific-conflict/ (Accessed on April 1st, 2020).

Beldomenico, P.M., S. Telfer, S. Gebert, L. Lukomski, M. Bennett, and M. Begon. 2008. "Poor Condition and Infection: A Vicious Circle in Natural Populations." *Proceedings of the Royal Society of London B: Biological Sciences* 275: 1753–9.

Boonstra, R., D. Hik, G.R. Singleton, and A. Tinnikov. 1998. "The Impact of Predator-Induced Stress on the Snowshoe Hare Cycle." *Ecological Monographs* 68: 371–94.

Bruers, S. 2018. "Thinking Critically About Wild Animal Suffering." Available at: www.youtube.com/watch?v=yQF6RSSLipM (Accessed on July 7th, 2020).

Clutton-Brock, T.H., and G.A. Parker. 1995. "Sexual Coercion in Animal Societies." *Animal Behaviour* 49: 1345–65.

Cohen, G.A. 2013. "Rescuing Conservatism: A Defense of Existing Value (All Souls Version)." In *Finding Oneself in the Other*. Ed. M. Otsuka (Princeton: Princeton University Press), pp. 143–74.

Creel, S., S. Liley, D. Christianson, and J.A. Winnie Jr. 2007. "Predation Risk Affects Reproductive Physiology and Demography of Elk." *Science* 315: 960.

Dawkins, R. 1995. *River Out of Eden: A Darwinian View of Life* (New York: Basic Books).

Delon, N., and D. Purves. 2018. "Wild Animal Suffering is Intractable." *Journal of Agricultural and Environmental Ethics* 31: 239–60.

Dworkin, G. 1976. "Autonomy and Behavior Control." *Hastings Center Report* 6: 23–8.

Dwyer, C.M. 2004. "How Has the Risk of Predation Shaped the Behavioural Responses of Sheep to Fear and Distress?" *Animal Welfare* 13: 269–81.

Dykes, L., and H. Orr. 2006. "Rabbit Analgesia: A Pragmatic Approach." *Veterinary Nursing Journal* 21: 13–5.

Elliot, R. 1982. "Faking Nature." *Inquiry* 25: 81–93.

Esvelt, K.M. 2018. "When Are We Obligated to Edit Wild Creatures?" Available at: www.leapsmag.com/when-are-we-obligated-to-edit-wild-creatures/ (Accessed on July 7th, 2020).

Farah, M.J. 2008. "Neuroethics and the Problem of Other Minds: Implications of Neuroscience for the Moral Status of Brain-Damaged Patients and Nonhuman Animals." *Neuroethics* 1: 9–18.

Francesconi, F., and O. Lupi. 2012. "Myiasis." *Clinical Microbiology Reviews* 25: 79–105.

Frankfurt, H.G. 1971. "Freedom of the Will and the Concept of a Person." *The Journal of Philosophy* 68: 5–20.

Getz, L.L. 1960. "A Population Study of the Vole, Microtus pennsylvanicus." *The American Midland Naturalist* 64: 392–405.

Hadley, J. 2015. *Animal Property Rights: A Theory of Habitat Rights for Wild Animals* (Lanham: Lexington Books).

Hardewig, I., H.O. Pörtner, and P. Dijk. 2004. "How Does the Cold Stenothermal Gadoid *Lota lota* Survive High Water Temperatures During Summer?" *Journal of Comparative Physiology* 174: 149–56.

Heithaus, M., A. Frid, and L. Dill. 2002. "Shark-Inflicted Injury Frequencies, Escape Ability, and Habitat Use of Green and Loggerhead Turtles." *Marine Biology* 140: 229–36.

Hettinger, N. 2018. "Naturalness, Wild Animal Suffering, and Palmer on Laissez-Faire." *Les ateliers de l'éthique/The Ethics Forum* 13: 65–84.

Horta, O. 2010. "Debunking the Idyllic View of Natural Processes." *Télos* 17: 73–88.

Hume, D. 1993. *An Enquiry Concerning Human Understanding*. Ed. Eric Steinberg (Indianapolis: Hackett Publishing).

Jeschke, J.M., W. Gabriel, and H. Kokko. 2008. "r-Strategists/K-Strategists." *Encyclopedia of Ecology* 4: 3113–22.

Johannsen, K. 2017. "Animal Rights and the Problem of r-Strategists." *Ethical Theory and Moral Practice* 20: 333–45.

Johannsen, K. 2020. "To Assist or Not to Assist? Assessing the Potential Moral Costs of Humanitarian Intervention in Nature." *Environmental Values* 29: 29–45.

MacArthur, R.H., and E.O. Wilson. 1967. *The Theory of Island Biogeography* (Princeton: Princeton University Press).
Maris, V., and E. Huchard. 2018. "Interventionnisme et Faune Sauvage." *Les ateliers de l'éthique/The Ethics Forum* 13: 115–42.
Marquis, D. 1989. "Why Abortion is Immoral." *The Journal of Philosophy* 86: 183–202.
McKinney, F., and S. Evarts. 1998. "Sexual Coercion in Waterfowl and Other Birds." *Ornithological Monographs* 49: 163–95.
McNamara, J.M., and A.I. Houston. 1987. "Starvation and Predation as Factors Limiting Population Size." *Ecology* 68: 1515–9.
Mill, J.S. 1885. *Three Essays on Religion* (London: Longman, Green & Co.).
Milburn, J. 2017. "Nonhuman Animals as Property Holders: An Exploration of the Lockean Labour-Mixing Account." *Environmental Values* 26: 629–48.
Mitani, J.C. 1985. "Mating Behaviour of Male Orangutans in the Kutai Reserve." *Animal Behavior* 33: 392–402.
Pianka, E.R. 1970. "On r- and K-Selection." *The American Naturalist* 104: 592–7.
Reimchen, T.E. 1988. "Inefficient Predators and Prey Injuries in a Population of Giant Stickleback." *Canadian Journal of Zoology* 66: 2036–44.
Schlottmann, C., and J. Sebo. 2019. *Food, Animals, and the Environment* (New York: Routledge).
Shriver, A. 2006. "Minding Mammals." *Philosophical Psychology* 19: 433–42.
Sinclair, A.R.E., and P. Arcese. 1995. "Population Consequences of Predation-Sensitive Foraging: The Serengeti Wildebeest." *Ecology* 76: 882–91.
Singer, P. 1993. *Practical Ethics, 2nd Edition* (Cambridge: Cambridge University Press).
Sládek, J.V. 1881. "Birds Suffering From Cold." *Nature* 24: 165.
Stoddart, D.M. (ed). 1979. *Ecology of Small Mammals* (London: Chapman and Hall).
Tizard, I. 2008. "Sickness Behavior, Its Mechanisms and Significance." *Animal Health Research Reviews* 9: 87–99.
Vargas-Terán, M., H.C. Hofmann, and N.E. Tweddle. 2005. "Impact of Screwworm Eradication Programmes Using the Sterile Insect Technique." In *Sterile Insect Technique.* Eds. V.A. Dyck, J. Hendrichs, and A. Robinson (Dordrecht: Springer).
White, T.C.R. 2008. "The Role of Food, Weather and Climate in Limiting the Abundance of Animals" *Biological Reviews* 83: 227–48.

# 3

# A COLLECTIVE OBLIGATION TO INTERVENE

## 3.1 Introduction

I concluded in the last chapter that the positive view of nature is false. Naturalness is not a source of intrinsic value, and it is only a very limited source of extrinsic value (our fallibility confers some value upon it). Furthermore, nature is not idyllic. In fact, nature is quite the opposite of idyllic: it's a place where suffering and premature death are the norm, rather than the exception. Having acquired a sense of what nature is actually like for wild animals, it's time to turn our attention to the moral implications of wild animal suffering (WAS). More specifically, does WAS entail a duty to intervene in nature, and if so, how should we understand that duty? It seems obvious enough that considerations of beneficence support intervention and thus, that any duty to intervene would be, at least in part, a duty of beneficence. Perhaps other considerations support intervention as well, though, for example, considerations of justice?

Whether the duty to intervene is best understood as a duty of beneficence is addressed in the fourth section of this chapter. In that section, I compare the conventional, beneficence-based understanding to two alternative understandings. One possibility is that our duty to wild animals should be understood along egalitarian lines, that is, as a duty of distributive justice owed in light of the fact that wild animals' opportunities to flourish are much worse than our own.[1] Another possibility is that what we owe to wild animals is specifically the rectification of anthropogenic harms such as those caused by climate change.[2] Though I believe that considerations of beneficence, distributive justice, and rectificatory justice, are all applicable, I argue that pro-interventionists have good reason to focus on beneficence in particular. Considerations of beneficence, unlike considerations of justice, are neutral between (a) animal welfarism: the view that animals' interests matter intrinsically but also far less than similar human interests, and (b) the animal rights position: the view that animals' interests matter just as much as similar human interests and thus, warrant the protection rights afforded.[3] As such, beneficence-based arguments are a better basis upon which to build democratic support for large-scale interventions. What's more, I'll argue that rectificatory justice runs into an agency-related problem that significantly limits its requirements. Rescuing harmful animals, such as predators and r-strategists, transforms the harms they subsequently cause into anthropogenic harms and thus, rectificatory justice only supports rescuing K-strategist herbivores.[4]

Before discussing how the duty to intervene should be understood, I respond to a series of objections to the claim that we have a duty to intervene in natural processes at all. In Section 3.3, I respond to the claim that we lack positive duties to wild animals because we lack a special relationship with them,[5] to the claim that we lack duties of beneficence to wild animals because wild animals are competent to manage the dangers they face,[6] and to the claim that beneficence requires providing wild animals with disaster relief but does not require intervention in natural processes.[7]

## 3.2 Preliminary Matters

In this section, I make a couple of preliminary points. I begin by arguing that our duty to intervene (assuming we have one) is best understood as a collective duty. Next, I devote some space to discussing the non-identity problem, and argue that pro-interventionists should not be troubled by it.

The first preliminary point I'd like to make is that any duty to intervene in natural processes on a large scale, for example, to eliminate predation or the r-strategy, would have to be a collective, political obligation, rather than a personal obligation. There are a number of reasons for this. For one, a duty to intervene on a large scale would be rather demanding. As noted in the last chapter, WAS is pervasive and its causes are many. Though individuals can successfully carry out certain small-scale interventions without too much trouble, for example, assisting an injured animal one comes across, or using social media to raise awareness about WAS, interventions that promise to make a significant, enduring difference in the lives of large numbers of wild animals require more resources than most people possess on their own. What's more, adequately addressing the ecological risks of intervention requires more than just research and testing – it requires coordination. After all, the ecological effects of any large-scale intervention depend, in part, on what other large-scale interventions are being conducted. The possibility that an intervention conducted by one person might harmfully interact with the interventions pursued by another person, suggests that those people who do have the resources to conduct large-scale interventions on their own, still shouldn't. In summary, an institutional solution is needed to ensure that individuals pool their resources and that those resources are subsequently used to develop and implement a coordinated intervention plan.

There are a number of questions to ask about which sort of institutions are best suited to carry out large-scale interventions. For example, are large-scale interventions best carried out at the national level by state institutions? Or would some sort of international solution be necessary? I won't attempt to provide a comprehensive answer, but I will note that some degree of international cooperation seems important. As I noted above, large-scale interventions carry ecological risks. What's more, ecosystems aren't confined within state borders. Since ecosystems cross borders, interventions undertaken by one country have the potential to affect another country's environment. At the very least, a nation implementing an intervention should seek the permission of other nations that may be affected ecologically. Furthermore, international cooperation is needed in order to avoid the possibility that interventions conducted by one nation might harmfully interact with the interventions pursued by another nation.

The second preliminary matter worth discussing is that future animals are the primary beneficiaries of large scale intervention. As Derek Parfit demonstrated more than 30 years ago, the idea that we have obligations to future generations is complicated by the fact that much of what we collectively do affects not only the level of well-being in future states of affairs, but also the identity of those who populate them.[8] The problem is that it's hard to see how an action could prevent harm when it also seems to prevent its beneficiaries from coming into existence. Similarly, it's hard to see how an action necessary for a person's existence could cause harm to them, so long as their life is worth living.

For example, suppose the state were to implement policies that successfully dissuade people from buying and driving personal motor vehicles. To keep things simple, let's also suppose that such dissuasion is the only effective way to lower emissions from personal vehicles. The relevant policies would have a variety of effects other than just decreasing emissions. People who otherwise would be driving cars to work now end up taking buses, riding bikes, etc., which in turn affects the people they end up meeting, as well as the times at which they leave for and return from work. These effects, in turn, would likely have some effect on people's reproductive behavior. For some, the effect will be big, for example, ending up with a different partner than one would have otherwise ended up with. For many, the change will be far smaller, for example, having intercourse at slightly different times than one otherwise would have. But even a slight change is significant, as changing the time at which a child is conceived also changes the relevant combination of gametes. When a different sperm and/or egg cell end up combining, the child who's conceived is non-identical with the child who would otherwise have been conceived. Here's the rub: someone who would only have been born in the absence of the relevant policies can't benefit from them, since she never comes into existence. Nor, so long as her life is worth living, does it seem that leaving emissions from personal vehicles as they were could be harmful to her, since she wouldn't exist otherwise. The relevant comparison is between existence and non-existence, rather than between existence at a certain level of well-being and existence at another, higher level of well-being.

Like policies that affect transportation, large-scale intervention would certainly affect reproductive behavior in various ways, thus affecting the identities of those who end up being born. In light of that, it's worth asking whether non-identity poses a problem for the claim that we have a collective duty to assist wild animals. How can we have a collective duty to help wild animals when, because of non-identity, many of those we seek to aid cannot, so it would seem, be coherently aided? Here's what I think. As noted in Chapter 2, it's plausible that, because of the r-strategy, most wild animal individuals do not live lives worth living. And for wild animals whose lives would not be worth living, it's certainly beneficial to them that other, better off animals, come into existence instead. But I don't think it's necessary to rely on the unproven, empirical claim that r-strategists' lives aren't worth living. After all, the non-identity problem is really just a theoretical puzzle. Though it might be difficult to understand how an altruistic but identity-affecting action could succeed in preventing harm to future people, or how a malicious (or simply negligent) but identity-affecting action could succeed in causing harm to future people, we have good reason to treat such actions as merely puzzling phenomena, thereby taking it for granted that they are in fact harmful or beneficial. It would be very counter-intuitive to accept that, so long as future generations live lives worth living, we should be indifferent toward the outcomes our identity-affecting behavior produces for them. Resolving the puzzle may simply require discovering previously unnoticed types of harm,[9] or it may require developing more sophisticated accounts of what it means to harm someone. Alternatively, we might take a consequentialist approach and accept that identity-affecting actions cannot either cause or prevent harm to the future people whose identities they affect (so long as their lives are worth living), but maintain that we have impersonal moral reasons to bring about states of affairs that contain higher levels of well-being, regardless of who may end up populating them.[10] Either way, we should accept that our behavior has morally significant consequences for future generations, even when those consequences are identity-affecting.

## 3.3 Objections to a Collective Duty to Intervene

### *3.3.1 An Outline of the Objections*

In this section, I'll respond to a series of objections to the claim that we have a collective, humanitarian duty to intervene in natural processes. According to the first objection, it's false that we generally owe positive duties to wild animals. In Clare Palmer's view, whether we owe positive duties to an animal depends on whether we share a special relationship with her, as well as on whether we're causally responsible for any harms that have or are likely to befall her.[11]

According to the second objection, we don't owe positive duties to wild animals because wild animals are not moral agents and are thus incapable of violating each other's rights. So long as no rights violation occurs when animals suffer natural harms, Tom Regan maintains that intervention is not morally required.[12] Furthermore, Regan responds to the objection that intervention is specifically a requirement of *beneficence*. His reply is that wild animals are competent to manage the dangers they face, and that competence is sufficient to block duties of beneficence.[13]

According to the last objection, we do indeed have some positive obligations to wild animals, but those obligations are not so extensive as to require large-scale interventions in natural processes. More specifically, John Hadley maintains that we only owe disaster relief to wild animals, and not the equivalent of development aid.[14] Though assisting animals affected by forest fires, earthquakes, and other natural disasters is something we can collectively accomplish without excessive cost, large-scale intervention in natural processes, such as the predator–prey relationship or wild animal reproduction, is allegedly too demanding for us to have even a collective obligation to do.[15]

An additional objection that's commonly made against a collective duty to intervene is the fallibility objection. It maintains that large-scale interventions in natural processes, even beneficent ones, risk ecological catastrophe and are thus, impermissible.[16] In Chapter 2, I devoted some space to discussing fallibility. More specifically, I argued that nature is only a source of value in so far as our fallibility makes it so. Though nature does not contain a high level of welfare, it's still better as it is, than as it would be in a scenario where human intervention inadvertently causes more harm than benefit. Of course, we also noted that the value nature acquires from fallibility recedes when enough research and testing has been done. It follows that fallibility is best understood as a constraint on justified intervention, rather than as a consideration that strictly bars it. Section 4.3 of Chapter 4 is devoted to further defending the view that fallibility is best understood as a constraint.

### 3.3.2 Relationships and Positive Duties

Let's begin with Palmer's objection. According to Palmer, positive duties, with the exception of duties of rectification, are relational. We owe positive duties, such as duties of care, to those with whom we share certain special relationships, for example, family members, friends, etc., but we lack positive duties when the relevant relationships are absent. Palmer maintains that only domesticated animals stand in a sufficiently close relationship with us to be owed such positive duties. Human beings control domesticated animals' breeding and thus, we are responsible for both their existence and the fact that they're dependent upon us to meet their needs. In Palmer's view, it's because we're causally responsible for domesticated animals' dependency that we in turn have a moral responsibility to satisfy their needs. Since wild animals were not bred by us, we allegedly don't owe them positive duties.[17]

The main problem with Palmer's view is that consistently applying it requires that we deny not only duties of care, but also duties of beneficence, to human beings with whom we lack any special relationship.[18] Though it's obvious that strangers don't owe each other the same extensive positive duties that parents owe their own children, we normally think we owe strangers certain duties of beneficence when they're in need and when we're in a position to help at little cost to ourselves, for example, checking on someone who has fainted and calling for help if they need it. Nor are such duties restricted to co-citizens or to those who reside in the same nation. When distant, foreign strangers are afflicted with famines, droughts, hurricanes, etc., we normally maintain that humanitarian assistance is owed.

For her part, Palmer is aware that her relational approach is in tension with the claim that we owe duties of humanitarian assistance to distant humans, and she recognizes that this tension is problematic. In an attempt to address it, she claims that duties of humanitarian assistance are also grounded in relationships, and not merely in the fact that distant, foreign strangers possess moral status. Though they do not share with us either family membership, friendship, community membership, co-citizenship, or even residence within the same nation, they are part of a global network of relations that wild animals don't participate in. Drawing on Leslie Pickering Francis and Richard Norman,[19] the relations Palmer highlights are "mutually recognized communication, the ability of humans to justify themselves to others, reciprocity in economic relations, mutual cooperation, [and] the joint organization of political and other institutions,"[20] among other things. It is because distant human beings take part in this global network of relations that we allegedly owe duties of assistance to them, but not to wild animals.[21] Upon inspection, though, Palmer's reply is insufficient to include all of the human beings we owe duties of assistance to. After all, not all human beings are capable of justifying themselves to others, of reciprocally participating in economic relations, or of participating in the organization of institutions. Palmer's relational account leaves out, for example, some human beings with severe cognitive disabilities.[22] If we're to generally include human beings within the scope of those to whom we owe duties of assistance, then we must allow that the possession of intrinsic moral status is enough to ground said duties on its own. But once we allow that moral status is enough, consistency demands that duties of assistance be extended to wild animals, too.

I mentioned above that Palmer's relational understanding of positive obligations permits one exception – duties of rectification. Duties of rectification are owed by an agent when they fail to satisfy their negative duty to refrain from causing harm, and since Palmer allows that we at least owe negative duties to wild animals, it follows that we owe them duties of rectification in cases where we've harmed them.[23] Allowing that we owe wild animals duties of rectification may not appear especially significant at first glance, but upon inspection, it opens the door to a fairly extensive level of intervention in the wild. Of particular note is the possibility that we owe rectification to wild animals that have been harmed by, or made more vulnerable because of, anthropogenic climate change.[24] Questions about the extent to which we owe rectification to wild animals will be addressed in section 3.4.

### *3.3.3 Natural Harms and Rights Violations*

Let's turn to the second objection against the claim that we have a humanitarian duty to intervene in natural processes. According to Tom Regan, respect for their rights requires that we refrain from harming wild animals, but it does not entail protecting them against natural harms. The reason, claims Regan, is that natural harms do not involve rights violations. Regan maintains that in order for a harm to also be a rights violation, it must be the case that the harm is culpably caused by a moral agent, that is, caused by a being who is morally responsible for, and thus to be blamed for, the harm. Human hunters, for example, violate the rights of the animals they kill. Human hunters, like other human beings, have the ability to reflect on and act upon moral reasons, and when they fail to do what's morally required of them, they are (under normal conditions) blameworthy. As a result, it's also appropriate to interfere with hunting by, for example, criminalizing it and imposing penalties. In contrast, natural causes of harm, such as predators, are not capable of violating wild animals' rights. Predators do not possess the cognitive abilities necessary to be held morally responsible for their behavior and thus, they are not blameworthy when they kill their prey. Because predators, unlike human hunters, are not capable of violating rights, respect for wild animals' rights allegedly does not require us to prevent predation.[25]

Though Regan focuses on predation specifically, his argument applies to many sources of suffering, the r-strategy included. After all, animals who give birth to large numbers of uncared-for offspring aren't morally responsible for their reproductive or parental behavior. If predators' lack of moral agency entails that we're not obligated to intervene in the predator–prey relationship, then presumably r-strategists' lack of moral agency entails that we're not obligated to interfere with their reproduction. However, moral agency (or rather its absence) does not have the implications Regan claims it does. More specifically, individuals who are threatened often have a claim to assistance even when the threat is not an agent. For example, we'd normally acknowledge that someone threatened by, for example, falling debris, has a claim to our assistance. Though I'm not required to help if doing so would necessitate placing myself in significant danger, I ought to assist so long as I'm in a position to do so at little cost to myself. What's more, we normally allow that individuals who are in need have a claim to our assistance, even when the cause of their need is not an agent, for example, those who are in need because of a natural disaster have a claim to humanitarian assistance. Similarly, then, the fact that predators aren't moral agents is no reason to think that the prey they threaten have no claim to assistance,[26] and the fact that r-strategist parents aren't moral agents is no reason to think that their offspring have no claim to assistance (even if the only way to feasibly help them is by preventing them from coming into existence in the first place). The main problem with Regan's argument is similar to the main problem with Palmer's. In both cases, consistent application of their views would require giving up the claim that we generally owe duties of assistance to human beings.

### *3.3.4 Competence and Beneficence*

In the 2004 edition of *The Case for Animal Rights*, Regan addresses the objection that we owe duties of assistance to wild animals, and that satisfying those duties would require intervening in natural processes such as the predator–prey relationship. Though Regan acknowledges that we owe duties of assistance to each other, he denies that beneficence requires intervening in natural processes. The reason he gives is that such intervention would be impermissibly paternalistic.[27] In Regan's view, it is in wild animals' interest to be allowed maximum personal liberty. Of particular importance is the ability to pursue one's interests one's own way. Regan maintains that paternalism, though well-intentioned, is often impermissible because it deprives animals of this ability. He uses the example of a captive wolf to illustrate. A captive wolf may have her desire for food more reliably satisfied than a free wolf, but she's also deprived of the satisfaction associated with securing the food for herself in her own way, for example, the satisfaction associated with cooperating with her pack, chasing her prey, etc.[28]

There are a number of problems with Regan's treatment of beneficence. First, his claim that beneficent interventions in the wild are *paternalistic* is not, generally speaking, true. To be paternalistic, an intervention must restrict a being's liberty for her own good. With respect to the above example, keeping a wolf captive is paternalistic so long as it's done for her own sake. The captor's justification would have to be that captivity is good for the wolf because it, for example, ensures that she's sheltered, fed regularly, etc. (the captor may, of course, be mistaken about what's in the wolf's interest, but that's beside the point). If the wolf is being kept captive in order to prevent her from killing prey, however, then her captivity is not paternalistic, as her liberty is being restricted for the sake of her prey's good, rather than her own good. Were the wolf a moral agent, the reason for her captivity would be the same as the reason for imprisoning a violent criminal – we'd be restricting her liberty in order to prevent her from violating the rights of others. But though Regan is right to note that predators cannot violate rights, they can (and do) cause harm, and restricting their liberty to prevent them from harming others is not paternalistic. Similarly, interfering with r-strategists' reproduction is not paternalistic. For it to be paternalistic, the interference would have to be for the sake of those who are being interfered with, but it isn't for the sake of prospective r-strategist parents that their reproductive liberty is interfered with. It's for the sake the offspring they would potentially have.

Second, even if beneficent intervention is sometimes paternalistic, it's not obvious that paternalistic interference is generally problematic in the case of wild animals. The reason is that the cognitive differences that exist between most adult human beings and most animals, affect the appropriateness of paternalistic interference. An example Regan himself provides is when a sick or injured animal is unable to understand the benefits of a veterinary procedure and thus, resists those who would have her undergo it.[29] Just as with children, situations like this call for a paternalistic response. In cases where we're dealing with adult human beings, however, paternalism isn't usually justified. For example, forcing an adult human to undergo a medical treatment is usually inappropriate, even when choosing the treatment is in their interest. The reason it's appropriate to coerce a sick animal, but not a sick adult human being, into undergoing a medically beneficial treatment, is because the capacity for critical reflection creates an important presumption against paternalistic intervention. More specifically, a person's awareness of and ability to reflect upon a conflict between their preferences and their objective interests, such as their medical interests, supplies a presumption in favor of respecting their preferences when they choose to act on them anyway. The reason is that what one chooses in light of critical reflection plays an important role in determining one's good. For example, though it might be in a patient's interest to choose a medically beneficial procedure, if, after considerable critical reflection, they choose to decline the procedure (on religious grounds, say), it is in their interest for the decision to be respected. In so far as animals are unable to possess such awareness or engage in such reflection, however, then there's no good reason to prioritize their preferences over their objective interests when the two conflict. This is why it's appropriate to coerce an animal into undergoing a medically beneficial procedure, but not appropriate to coerce an adult human being.

In reply, I think Regan would claim that wild animals' preferences, and their objective interests, don't typically come into conflict. He maintains that wild animals are generally competent to manage the dangers they face, and that when they possess enough competence, having the freedom to purse their preferences is also in their interest.[30] Consider again the captive wolf example. Regan notes that though a captive wolf has her dietary needs more reliably satisfied than a free wolf, captivity is nonetheless not in her interest, as captivity deprives her of the satisfaction associated with chasing prey, cooperating with the pack, etc.[31] Were the wolf seriously ill, though – too ill to secure her own food, keep up with the pack, or defend herself – then a temporary period of captivity at a wildlife hospital is likely in her interest. She may not be able to experience the satisfaction associated with securing her own food while there, but since she's too ill to secure her own food anyway, it's only in her interest to be released after she's recovered. This suggests an important difference between captivity and veterinary procedures. Whereas the captive wolf (when healthy) is competent to secure her own food, sick animals are not competent to cure their own ailments. This is why it's in the wolf's interest to be released, but not in a sick animal's interest to be left alone.

Regan thinks that wild animals, including prey, are generally quite competent. In a telling passage, he writes that

> our ruling obligation with regard to wild animals is *to let them be*, an obligation grounded in a recognition of their general competence to get on with the business of living, a competence that we find among members of both predator and prey species. After all, if members of prey species, including the young, were unable to survive without our assistance, then there would be no prey species. And the same applies to predators.[32]

Were it true that wild animals are generally competent, Regan's objection to intervening in the wild would have some force against paternalistic interventions. It would also have some force against non-paternalistic interventions. For example, if r-strategist infants were generally competent to manage the dangers they face, then we would have far less reason to interfere with their parents' reproduction. Preventing r-strategists from giving birth to doomed offspring presupposes that, in fact, their offspring are doomed. Unfortunately, and contrary to what Regan claims, the fact that a species exists doesn't mean that its individual members are competent. As we already know, most members of (sentient) r-strategist species die very painfully at a very young age. In fact, there's a stark conflict between the survival of an r-strategist species and the survival of its individual members, since increasing the quantity of offspring produced decreases the amount of parental energy available for each individual infant. And since the vast majority of wild animal individuals are r-strategists, it's false that wild animals are generally competent.

A second point worth noting is that competent risk management is a kind of pragmatic second best – a concession to non-ideal circumstances. It would be better if flourishing did not necessitate enduring risks at all. Of course, in cases where facing a risk is truly necessary for flourishing, our best option is to endure the risk. By and large, however, the necessity of facing a risk is contingent upon one's circumstances, and circumstances can often be changed. For example, letting children play outside, walk home from school, explore their neighborhood, etc., is important for their flourishing, even though doing so involves various risks. Because these activities are important, parents should permit their children to face the associated risks on their own once a certain level of competence has been achieved. But some neighborhoods are riskier than others, for example, some have higher crime rates, more vehicular traffic, etc. Living in a safer neighborhood, where one's children can flourish while facing fewer risks, is ideal. After all, children who are good at crossing the street are still at a higher risk of being hit by a car, if they live on a busy street. And children who are good at avoiding ruffians are still at a higher risk of being assaulted, if they live in a rough neighborhood.

Though letting children manage the risks of their neighborhood is preferable to perpetually sheltering them, reducing the extent to which outdoor activities present risks in the first place (by, e.g., leaving a bad neighborhood) is even better. Of course, even the safest neighborhoods contain some risks, and the parents who live in them should still prepare their children. The point is just that, all things equal, the absence of risk is better than risk management.

As I noted earlier, wild animals are not, generally speaking, competent to manage the dangers they face. But some wild animals are. K-strategists are far better at surviving than r-strategists are, so it's fair to say that they're somewhat competent. In light of that competence, I don't think it's in a wild K-strategist animal's interests to be kept captive in a zoo. Being in a zoo ensures access to food and protection from predators, but it also deprives an animal of important liberties. Still, though a life of confinement and dependency is arguably worse for wild K-strategists than facing the risks of life in the wild, life in a wilderness full of risks is worse than life in a wilderness that's safer. Consider predation once more. As noted in Chapter 2, predation doesn't just cause injury and death – it's also a significant source of stress for prey animals, and the threat of predation often prevents prey from foraging for food, thereby causing many of them to go hungry. In light of the suffering associated with predation, life in an ecosystem where one is not at risk of being attacked by predators is, all things equal, better than life in an ecosystem where one is at risk. Though they're not the norm, it's encouraging to observe that ecosystems containing little predation exist, for example, on certain remote islands, and that the prey animals who live in them are more relaxed than their predator-plagued relatives – a phenomenon known as 'island tameness.'[33] We're not presently able to safely create predator-free ecosystems ourselves, but we may, with enough research, be able to do so one day through gene editing.[34] Furthermore, we should be able to mitigate some of the other risks associated with life in the wild, for example, the risk of going hungry. It may have initially appeared to be in the interest of specifically *competent* wild animals to face such risks themselves, but as we now know, risk management is only preferable to sheltered dependency. Even better is when one's circumstances permit one to flourish without facing certain risks at all. The upshot is that competence, in so far as it exists in the wild, only undermines liberty-restricting interventions. Interventions that don't involve perpetual interference with wild animals' liberties, such as CRISPR gene drives, can be appropriately used to assist K-strategists, not just r-strategists.

### 3.3.5 Assistance and Development Aid

Let's turn to the final objection of interest to us. John Hadley, unlike Palmer and Regan, does not deny that we owe duties of assistance to wild animals. In fact, he argues that, in at least some circumstances, we're morally required to assist wild animals who've been negatively affected by natural disasters such as droughts, forest fires, etc.[35] What Hadley denies is that our commitment to assistance requires us to intervene in the predator–prey relationship (he's presumably against intervening in other natural processes, too, e.g., wild animal reproduction). Hadley offers a couple of different reasons in support of his view. The first is that intervening in the predator–prey relationship would be very costly and thus, that it cannot be one of our duties of assistance, since duties of assistance are only owed when they can be discharged at little cost to one's self.[36] The second is that preventing predation is not, properly speaking, a matter of assistance. Duties of assistance specifically require that we help others in need, and prey animals are allegedly not in need until they're actually being attacked. In Hadley's view, preventing predation is more akin to development aid than to assistance, and whether we owe development aid to wild animals is a separate question.[37] Though Hadley doesn't explicitly say it, a duty to provide development aid would be a duty of distributive justice, rather than a duty of humanitarian assistance.

The possibility that we owe duties of distributive justice to wild animals will be discussed in the next section. In this section, I'll restrict myself to addressing Hadley's claim that duties of assistance do not extend to the predator–prey relationship. First, his claim that a duty to intervene in the predator–prey relationship would be too demanding, is problematic on a couple of grounds. As we noted earlier, any duty to intervene in nature on a large scale would of course be too demanding for most people to fulfill individually – that's precisely why it would have to be a collective duty. Collective duties discharged through our institutions are not especially onerous. For most people, they only involve paying somewhat higher taxes. Of course, we don't currently have institutions that systematically aid wild animals on a large scale and thus, advocacy is presently needed.[38] Since doing what's necessary to ensure that the right institutions are put in place has the potential to be quite demanding, any duty to advocate for institutional change is constrained in the same manner that duties of assistance generally are. More specifically, one is required to advocate for change only in so far as one can do so at little cost to one's self.

The second problem with the claim that intervention is too costly is that it assumes the only interventions available are those which require perpetual labor from large numbers of people. For example, Hadley points out that 'policing' nature in an effort to prevent predation, would be quite costly. Though Hadley's right to note that extensive labor would be necessary to systematically separate predators from prey using conventional methods, such as fences and 'police officers,' such methods aren't our only options. In particular, gene editing promises to be rather cost-effective. The initial research needed to reprogram predators to eat plants,[39] or to develop plants suitable for predators to eat,[40] will obviously involve some costs, but once the relevant gene drives have been developed and administered, it isn't necessary to sustain them. Unlike a police system, which requires constant maintenance, gene drives only involve initial costs.[41] What's more, the gene editing method scientists currently use – CRISPR – is far cheaper than earlier forms of gene editing,[42] and it's always possible (perhaps likely) that we'll develop even cheaper gene editing technology in the future.

In addition to claiming that intervention in the predator–prey relationship is too costly to be a duty of assistance, Hadley also claims that, even if it was required, it would not be a requirement of *assistance*. Instead, he maintains that it would be a form of development aid, and that whether we owe wild animals development aid is a different question than whether we owe them assistance. I have two points to make in reply. First, it isn't true that there's a morally significant difference between preventing harm and assisting. This is because a duty to prevent harm, and a duty to assist those in need, are both duties of beneficence. To be fair, a duty to *refrain from causing harm* is a duty of non-maleficence and is thus, different in kind from a duty of assistance, but assisting others and preventing harm to them are both grounded in the same moral value (beneficence).[43] As such, if we have a duty to assist wild animals when we can do so at little cost to ourselves, then we presumably also have a duty to prevent harm from befalling them when we can do so at little cost to ourselves.

Second, applying the distinction between assistance and development aid in the manner Hadley does presupposes that, at least under normal conditions, wild animals are not in a state of need. It presupposes that the sort of life-threatening dangers wild animals face when natural disasters strike, namely dangers they can't cope with on their own, are out of the ordinary. In other words, Hadley presupposes that wild animals are generally successful at managing the risks they face. Of course, Hadley isn't the only philosopher who assumes that most wild animals competently manage the risks they face. Regan makes the same assumption. In fact, a number of philosophers, including Regan, go one step further than Hadley. Their view is that most wild animals not only competently manage the risks of life in the wild, but that so long as animals are competent to manage a risk, intervening in order to protect them from that risk is contrary to their interests.[44] In other words, these philosophers maintain that non-interference is necessary for wild animals' flourishing. As we already noted in our discussion of Regan, though, there are two problems with this type of argument. First, competent risk management is a concession to non-ideal circumstances. Facing a necessary risk is better for animals than sheltered dependency is, but removing the need to face that risk is better yet. Second, most wild animals do not competently manage the risks they face. Instead, most of them die painful deaths shortly after birth. In light of this fact, it's fair to say that dire need is presently the norm in nature, rather than the exception.

## 3.4 Understanding Our Collective Obligation: Justice or Beneficence?

In the previous section, I hope to have successfully defended the claim that WAS justifies a collective duty to intervene in natural processes. In this section, I consider the question of how that duty should be classified. Thus far, I've assumed that our collective duty is specifically a duty of beneficence. However, beneficence isn't the only ground we have for intervention. As Alasdair Cochrane argues, cosmopolitan justice strongly militates in favor of intervention, too.[45] After all, the core intuition driving cosmopolitanism is that the circumstances of one's birth, including one's nationality, are morally arbitrary. Whether one happens to be born into a rich family in a wealthy nation, or into a poor family in an impoverished nation, is a matter of sheer luck for the person who's born. Accordingly, cosmopolitans maintain that the inequality between the global rich and the global poor is unfair, in so far as it's traceable to nationality. But, if we judge it unfair for the circumstances of birth to determine one's (relative) life prospects, doesn't that judgment extend to species membership, too? The species one happens to be born into is just as circumstantial as the nation one's born in. And if inequalities traceable to species membership are unfair, then wild animals are in a tremendously unfair situation. On average, wild animals are far worse off than human beings. Even a wild animal born into a K-strategist species is worse off than most people born into a human nation are (including those born into fairly poor nations). In light of these inequalities, do we not have reasons of distributive justice to intervene in nature?

In addition, Clare Palmer, as noted above, has argued that we owe rectification to wild animals. She highlights various anthropogenic harms that she thinks ground duties of rectification, but the most significant harms are likely those associated with anthropogenic climate change. In her words,

> Changing precipitation patterns and intensity, rising temperatures, warming of the upper oceans, Arctic ice melt, sea level rise, heat waves, the shifting of habitat types, will all impact wild animal populations. While some populations will be able to take advantage of the changing conditions to expand and grow, others will be severely affected, including many already threatened and endangered species.[46]

Considering how pervasive the effects of climate change both are and will be, it seems that not only distributive, but also rectificatory justice, militate in favor of a considerable level of intervention in nature. The forms of rectificatory intervention Palmer considers are rescue and rehabilitation, habitat restoration, and assisted migration.[47]

I think it's true that considerations of both distributive and rectificatory justice support intervention in the wild. However, I also think that we should focus specifically on beneficence, and the interventions that beneficence supports. The reason is that there are a number of theoretical and practical problems with applying principles of distributive justice and principles of rectificatory justice to wild animals, problems that don't attend the application of beneficence. I'll start with distributive justice.

The first major issue with grounding intervention in an appeal to unfair inequality, is that the unfairness of inequality is contingent upon whether animals are our moral equals. More specifically, if animals are our moral equals, then inequality (in the distribution of life chances) between them and us is unfair, but if animals are morally inferior to us, then inequality is not unfair. Though I myself adhere to the view that animals are our moral equals, I also recognize that my view is a minority view, and that its status as a minority view has political implications. Before, we get into those implications, though, let's consider why moral equality affects the moral significance of distributive inequality. As I understand it, to say that animals are our moral equals is to say that their interests matter just as much as our interests.[48] The view that all sentient beings' interests warrant equal consideration grounds the view that animals possess the same basic moral rights we do, for example, the right to life, the right not to have one's liberty arbitrarily restricted, etc., and it suggests that a state of affairs in which human interests are typically satisfied but wild animals' interests are not, is unfair. After all, if wild animals' interests matter just as much as ours, then it's at least highly regrettable that they, unlike us, have very little opportunity to satisfy those interests. And if there are low-cost, feasible actions we can take to ensure that they do have the opportunity to satisfy their interests, or at least low-cost actions we can take to make the attainment of that opportunity feasible in the future, then the inequality between us and them isn't just regrettable: it's also wrong. But if, as most people believe, animals' interests matter far less than our own, then there's nothing unfair about the inequality between us and them.

Of course, one should still regret the fact that wild animals' lives tend to be miserable, even if one doesn't believe that animals are our moral equals. Though there was a time when some defended the view that animals lack intrinsic moral status, and that our animal-related duties are strictly indirect duties owed to the human beings we might one day harm if we cultivate cruelty in ourselves,[49] almost no one holds this view now. Instead, most people are animal welfarists.[50] Animal welfarism maintains that animals' interests matter in and of themselves, but that human interests trump animal interests whenever there's a conflict. A claim characteristically made by proponents of this view is that animal suffering is morally bad, but that it's also justified when necessary to fulfill some human interest, for example, an interest in meat consumption or in medication. As such, animal welfarists maintain that unnecessary sources of suffering ought to be eliminated, for example, that farm animals should be slaughtered as painlessly as possible, that cage sizes should be increased, etc.

From an animal welfarist's perspective, there's nothing unfair about the fact that wild animals have a lesser opportunity to fulfill their interests than us. However, animal welfarism does recognize that wild animals' suffering is morally bad. In light of that recognition, duties of not only non-maleficence, but also beneficence, are compatible with it. If, for example, one observed someone kicking a duck for no reason, then it would be consistent with animal welfarism to recognize that the observer has a duty to intervene for the sake of preventing harm to the duck, so long as they can do so without harming themselves or the kicker in the process, for example, by yelling at the kicker to stop. This, then, is the first main reason for focusing on beneficence rather than distributive justice. Unlike distributive justice, considerations of beneficence apply regardless of whether animals are our moral equals and thus, the soundness of a beneficence-based case for intervention does not require that the animal rights position be true.

It might strike the reader as strange that I want the case for intervention to avoid presupposing the animal rights position. As I mentioned before, I myself hold the animal rights position, and I myself accept that considerations of distributive justice support intervening in the wild. So, why am I keen to argue that we should focus on considerations that are consistent with animal welfarism? I have a few reasons. First, it's always possible that the animal rights view is false. Though the animal rights view seems to me to be well-justified – enough so that I'm reasonably confident in its truth – it would be epistemically immodest not to acknowledge that I could be wrong. Given this possibility, it makes sense to at least bolster the case for intervention with considerations that don't depend on the animal rights position. But this isn't the only reason for my interest in neutral considerations. A second reason is that neutral considerations have far more political traction than non-neutral ones. After all, the animal rights position, as I already noted, is a minority view. Since most people are animal welfarists, motivating political support for intervention in nature will require appealing to considerations that animal welfarists can accept.[51]

It's worth emphasizing that political traction is about more than just strategy. It's also about legitimacy. Implementing institutional programs sufficient to aid wild animals on a large scale will require securing public funds through taxation, and the procurement of such funds requires public support. If a reasonable justification can't be given to those who are to be taxed, that is, a justification that they can reasonably be expected to accept, then it will be very difficult to raise the public support necessary to democratically implement large-scale interventions. And intervention should democratic if it's to be legitimate.

It might be objected that my case for neutral considerations incorrectly presupposes that disagreement about the scope of moral equality is part of, rather than prior to, reasonable pluralism. My objector maintains that in order for a nation's government to be legitimate, it's laws and policies must respect and protect the basic moral rights of residents, for example, their rights to life and liberty. A government whose military goes about arbitrarily violating residents' rights, or whose laws permit institutions that routinely violate rights, for example, institutionalized slavery, loses its claim to legitimacy *even if* it enjoys popular support. But if it turns out that animals possess basic moral rights, then presumably their rights must be respected too, in order for the state to be legitimate.[52] From this perspective, catering to animal welfarists seems wrongheaded. We should demand that animals' equal moral status be recognized, and that the duties associated with it be fulfilled.

Though I'm sympathetic to the claim that failing to respect animals' rights undermines a state's legitimacy, I also think it's important to keep in mind that our duty to intervene in the wild is specifically a *positive* duty to *wild* animals, and that failing to fulfill duties of this sort, though morally problematic, does not undermine a state's claim to legitimacy. One reason for thinking this is that, for the most part, negative duties are stricter than positive duties. Another reason is that a positive duty to a wild animal is weaker than an equivalent positive duty to a domesticated animal, since we bring domesticated animals into existence and are thus, responsible for them. As such, failing to satisfy a *positive* duty to a *wild* animal is, in two important respects, different from failing to satisfy a *negative* duty to a *domesticated* animal. It's far more plausible to claim that failing to satisfy negative duties to domesticated animals undermines legitimacy, than it is to claim that failing to satisfy positive duties to wild animals does. Consider a human analogy – positive duties owed to the impoverished citizens of third world nations. Though the inequality between first world and third world citizens is certainly unfair, is the legitimacy of a democratically elected, first world nation's government undermined if it fails to fund a considerable amount of aid? Put another way, would violent resistance against that government be justified, and would military interference from other governments become appropriate? Presumably not. In fact, I think most of us would agree that lesser forms of coercion, such as political or economic sanctions, are still inappropriate. In contrast, if a democratically elected government decided to massacre an ethnic minority, or to legalize their enslavement, we would normally say that this government has lost its claim to legitimacy, regardless of whether it enjoys popular support.

The above contrast suggests an important, political difference between failing to intervene in the wild, and failing to prohibit the actions and institutions animal rights advocates traditionally criticize. On the assumption that animals are our moral equals, it may very well be true that states must, for example, ban factory farming in order to be truly legitimate. After all, factory farming is in many respects analogous to slavery, the main difference being that factory farm animals, unlike slaves, are routinely killed by their owners. If this is right, then, from the standpoint of legitimacy, it doesn't matter whether factory farms enjoy popular support, and coercively pressuring a government to get rid of them, even through the use of violence, may sometimes be justified. But failing to intervene in the wild is not analogous to failing to ban factory farms (or to failing to ban hunting, even). Large-scale intervention in the wild, to be legitimate, must be democratic, and democratic implementation requires popular support. What's more, garnering popular support requires that the arguments employed make use of shared reasons. I hope that one day the animal rights position will be accepted by the majority of people, but until then, garnering support for intervention will require making arguments that animal welfarists can accept, too.

Thus far, I've argued that it's better to ground large-scale intervention in considerations of beneficence than in considerations of distributive justice, because a beneficence-based case doesn't presuppose the animal rights position. But what about considerations of rectificatory justice? What are we to make of them? The basic idea behind rectificatory justice is that if someone has a right not to be harmed, but we harm them nonetheless, our failure generates a secondary duty to rectify the harm we caused, for example, to financially compensate someone for their medical expenses, for damaged property, etc. Of course, if a being doesn't have a right not to be harmed, then she cannot have a right to rectification in the event that she is harmed. We should thus note that, like distributive justice, applying principles of rectificatory justice to wild animals presupposes the animal rights position. Though it might seem that rectificatory justice should have as little political traction as distributive justice, I'm not so sure. Interventions supported by rectificatory justice coincide with environmentalists' intuitions much better than interventions supported by beneficence or distributive justice. After all, harms associated with natural processes, such as predation or the r-strategy, are not anthropogenic. Since we're not responsible for causing such harms, we do not, from the perspective of rectificatory justice, have a duty to assist animals affected by them. Instead, our duty would be to assist animals who have been or will be, harmed by our behavior, for example, animals harmed by anthropogenic climate change. Unlike intervening in predation or the r-strategy, intervening to assist victims of climate change is often supported by environmentalist intuitions. Examples mentioned above include rescuing and rehabilitating negatively affected animals, as well as restoring their habitats. Support from environmentalists makes at least some rectificatory interventions politically feasible.

Rectificatory justice may have some role to play in determining which interventions ought to be funded. However, I think that role should be highly circumscribed. This is because rectificatory justice cannot consistently support saving animals who, once saved, will predictably go on to cause significant harms, namely r-strategist animals or predators. Though human beings aren't normally the cause of harms associated with predation or the r-strategy, we become causally responsible when we save predators' or r-strategists' lives. A person who, for example, rescues a sick predator on the verge of death, rehabilitates her, and then releases her into the wild, is causally and morally responsible for the injuries and deaths that predator subsequently causes. In other words, someone who rehabilitates a predator violates the rights of her prey.

One way to try and avoid this conclusion is to ascribe moral agency to predators. If predators are morally responsible for their behavior, then perhaps the rehabilitated predator, rather than the person who saves her, is blameworthy. Indeed, some animal ethicists have argued that some animals are, to some extent, moral agents, though the matter remains controversial.[53] But even if it were true that some predators are full moral agents, it isn't true that someone who saves them shares no moral responsibility for the prey they subsequently kill. Rescuing and rehabilitating an agential predator is similar to rescuing and rehabilitating a known serial killer. Though the serial killer (who, for the sake of the argument, we'll assume is not insane) is responsible for their behavior, the fact that they'll predictably strike again means that their rescuer bears some responsibility, since rescuing them enables them to kill once more. Nor does it matter if one rescues and rehabilitates a serial killer for purposes of rectification. Regardless of whether one finds the serial killer freezing to death in the cold, or whether one picks them up off the side of the road after accidentally hitting them with a car, a rescuer who knowingly lets a rehabilitated killer go free is blameworthy, and we should think the same of someone who rehabilitates an agential predator.

An additional problem is that ascribing agency to predators means that, from an animal rights perspective, the harms they commit qualify as rights violations. Recall that Tom Regan's initial response to the predation problem is that because predators are not moral agents, the harms they commit are not rights violations and thus, there is (allegedly) no duty of justice to prevent them from harming prey. Though Regan's response was not sufficient to deflect the claim that we have a duty to intervene in the predator–prey relationship, it's certainly true that the presence of rights violations would strengthen the case for intervention. As such, ascribing agency to predators, or to r-strategists for that matter, is in tension with maintaining the rectificatory justice approach's restricted focus on anthropogenic harms. If predators violate the rights of their prey, and if r-strategist parents violate the rights of their young, then intervention in natural processes is hard to consistently argue against.

## 3.5 Conclusion

In conclusion, this chapter has defended two main claims: (a) that WAS justifies a collective duty to intervene in nature, and (b) that pro-interventionists have good reason to focus on considerations of beneficence, rather than on considerations of distributive justice or rectificatory justice. With respect to claim (b), I argued that, in the animal case, distributive justice has far less political traction than beneficence does. I also argued that the scope of rectificatory justice is fairly narrow since it only supports saving K-strategist herbivores and not predators or r-strategists. With respect to claim (a), I proceeded by replying to a series of objections. In reply to Palmer's claim that we have no special relationship with wild animals, and to Regan's claim that wild animals cannot violate each other's rights, I argued that neither point has any bearing on whether beneficence requires intervention. In reply to Regan's later claim that wild animals are competent to manage the dangers they face, I first argued that his claim is largely false, since most individual wild animals are r-strategists, and since most r-strategists die painfully and prematurely. Next, I argued that even with respect to competent wild animals (K-strategists), intervention is justified. Though competence undermines paternalistic interventions, not all large-scale interventions infringe upon the liberty of those they assist. What's more, though competent risk management is better than sheltered dependency, removing the necessity of a risk is better than merely managing it. Finally, in reply to Hadley's claim that beneficence justifies disaster relief but not intervention in natural processes, I argued that he overestimates the costs of intervention, exaggerates the difference between providing assistance and preventing harm, and falsely assumes that most wild animals are competent to manage the dangers they face.

## Notes

1 Faria (2014), Horta (2016), and Cochrane (2018).
2 Pepper (2019), Palmer (2019), and Sebo (forthcoming).
3 Though I don't discuss it, equal consideration comes apart from rights on some views. For example, utilitarians often deny that morally significant beings have rights, but they don't deny that morally significant beings warrant equal consideration. For a utilitarian, equal consideration is just a matter of according like interests the same weight when deciding which course of action would maximize utility. Furthermore, one could deny that the interests of animals warrant equal consideration, and yet maintain that they warrant consideration sufficient to ground at least some rights.
4 To be fair, there are some conditions under which rectificatory justice supports rescuing predators and r-strategists. For example, it likely supports rescuing older, herbivorous r-strategists who are no longer able to reproduce. It may also be the case that rectificatory justice supports rescuing predators and r-strategists so long as measures are taken to prevent them from subsequently causing harm, e.g., sterilizing rescued r-strategists or keeping rescued predators in captivity. However, I expect that rescues accompanied by harm prevention measures are only justified (from the perspective of rectificatory justice) so long as the benefits to a rescued animal exceed the costs incurred.
5 Palmer (2010) pp. 84–95 and Palmer (2015).
6 Regan (2004) p. xxxvii.
7 Hadley (2006) pp. 449–50.
8 Parfit (1984) chapter 16.
9 See, for example, Shiffrin's discussion of non-comparative harms in Shiffrin (1999) pp. 120–35.
10 For a discussion, see Parfit (1984) chapters 17 and 19.
11 Palmer (2010) pp. 84–95.
12 Regan (2004) p. 357.
13 Regan (2004) p. xxxvii.
14 Hadley (2006) p. 450.
15 Hadley (2006) p. 450.
16 Singer (1975) pp. 238–39, Simmons (2009), and Ladwig (2015) pp. 297–9. For a recent, extensive discussion of the epistemic hurdles associated with large-scale intervention, see Delon and Purves (2018).
17 Palmer (2010) pp. 84–95 and Palmer (2015).
18 Donaldson and Kymlicka (2011) p. 162
19 Francis and Norman (1978).
20 Palmer (2010) p. 121.
21 Palmer (2010) pp. 119–24.
22 Catia Faria offers a compelling set of replies to Palmer in Faria (2015). See p. 214 in particular.
23 Palmer (2010) pp. 88–9.
24 Palmer (2019).
25 Regan (2004) p. 357. For his distinction between moral agents and moral patients, see Regan (2004) pp.151–6.

26 See Jamieson (1990) p. 351. See also Cowen (2003) p. 176 and Simmons (2009) pp. 19–20.
27 Regan (2004) p. xxxvii.
28 Regan (2004) pp. 91–2.
29 Regan (2004) p. 108.
30 Regan (2004) p. xxxvii.
31 Regan (2004) p. 92.
32 Regan (2004) p. xxxvii.
33 See, for example, Cooper et al. (2014).
34 I discuss gene editing in Chapter 5 of the present book. For work exploring the idea that we should use gene editing to phase out predation, see McMahan (2015) and Pearce (2015).
35 Hadley (2006).
36 Hadley (2006) p. 450.
37 Hadley (2006) p. 450.
38 It's for this reason that Chapter 6 of the present book is devoted to issues that pertain to advocating for humanitarian intervention in nature.
39 McMahan (2015) and Pearce (2015).
40 Johannsen (2017) p. 342.
41 Milburn (2019) pp. 23–4.
42 Gersbach (2014) pp. 1009–10 and Ledford (2015) p. 21.
43 See, for example, the discussion on beneficence in Beauchamp and Childress (2013) pp. 202–4.
44 See Regan (2004) p. xxxvii and pp. 91–2, Everett (2001) pp. 54–5, and Donaldson and Kymlicka (2011) pp. 179–87.
45 Cochrane (2018). See also Faria (2014) and Horta (2016).
46 Palmer (2019) p. 1.
47 Palmer (2019) pp. 7–14. For additional discussion and argumentation in support of the claim that the effects of climate change create duties of rectification to wild animals, see Pepper (2019) and Sebo (forthcoming).
48 See, for example, Singer (1975) pp. 5–6.
49 This view is normally associated with Immanuel Kant. For a critical discussion of it, see Regan (2004) pp. 174–85.
50 For a helpful description of animal welfarism that contrasts it with both rights-based and ecology-based understandings of animals' moral significance, see Donaldson and Kymlicka (2011) pp. 3–5.
51 Other animal ethicists have stressed that one should appeal to neutral considerations when arguing for moral and legal reform. See, for example, the discussion on internal vs. external inconsistency in O'Sullivan (2016) pp. 57–9 and pp. 63–5, and the discussion on cruelty in Milligan (2016).
52 For a defense of this view, see Cochrane (2018) chapter 6.
53 For an article that surveys the debate over whether some animals are moral agents, see Clement (2013).

## References

Beauchamp, T.L., and J.F. Childress. 2013. *Principles of Biomedical Ethics, 7th Edition* (New York: Oxford University Press).

Clement, G. 2013. "Animals and Moral Agency: The Recent Debate and Its Implications." *Journal of Animal Ethics* 3: 1–14.

Cochrane, A. 2018. *Sentientist Politics: A Theory of Global Inter-Species Justice* (Oxford: Oxford University Press).

Cooper Jr., W.E., R.A. Pyron, and T. Garland Jr. 2014. "Island Tameness: Living on Islands Reduces Flight Initiation Distance." *Proceedings of the Royal Society B: Biological Sciences* 281: 20133019.

Cowen, T. 2003. "Policing Nature." *Environmental Ethics* 25: 169–82.

Delon N., and D. Purves. 2018. "Wild Animal Suffering is Intractable." *Journal of Agricultural and Environmental Ethics* 31: 239–60.

Donaldson, S., and W. Kymlicka. 2011. *Zoopolis: A Political Theory of Animal Rights* (New York: Oxford University Press).

Everett, J. 2001. "Environmental Ethics, Animal Welfarism, and the Problem of Predation: A Bambi Lover's Respect for Nature." *Ethics and the Environment* 6: 42–67.

Faria, C. 2014. "Equality, Priority and Non-Human Animals." *Dilemata* 14: 225–36.

Faria, C. 2015. "Disentangling Obligations of Assistance: A Reply to Clare Palmer's 'Against the View That We Are Normally Required to Assist Wild Animals'." *Relations* 3: 21118.

Francis, L.P., and R. Norman. 1978. "Some Animals Are More Equal than Others." *Philosophy* 53: 507–27.

Gersbach, C.A. 2014. "Genome Engineering: The Next Genomic Revolution." *Nature Methods* 11: 1009–11.

Hadley, J. 2006. "The Duty to Aid Non-Human Animals in Dire Need." *Journal of Applied Philosophy* 23: 445–51.

Horta, O. 2016. "Egalitarianism and Animals." *Between the Species* 19: 108–44.

Jamieson, D. 1990. "Rights, Justice, and Duties to Provide Assistance: A Critique of Regan's Theory of Rights." *Ethics* 100: 349–62.

Johannsen, K. 2017. "Animal Rights and the Problem of r-Strategists." *Ethical Theory and Moral Practice* 20: 333–45.

Ladwig, B. 2015. "Against Wild Animal Sovereignty: An Interest-Based Critique of Zoopolis." *Journal of Political Philosophy* 23: 282–301.

Ledford, H. 2015. "CRISPR, the Disruptor." *Nature* 522: 20–4.

McMahan, J. 2015. "The Moral Problem of Predation." In *Philosophy Comes to Dinner: Arguments about the Ethics of Eating*. Eds. A. Chignell, T. Cuneo, and M. Halteman, (London: Routledge), pp. 268–94.

Milburn, J. 2019. "Sentientist Politics Gone Wild." In *Book Symposium: "Sentientist Politics: A Theory of Global Interspecies Justice."* By A. Cochrane, J. Milburn, and S. O'Sullivan, pp. 19–24 (*Politics and Animals*, Vol. 5).

Milligan, T. 2016. "Putting Pluralism First: Cruelty and the Animal Rights Discourse." In *The Political Turn in Animal Ethics*. Eds. R. Garner and S. O'Sullivan (London: Rowman & Littlefield), pp. 85–101.

O'Sullivan, S. 2016. "Animals and the Politics of Equality." In *The Political Turn in Animal Ethics*. Eds. R. Garner and S. O'Sullivan (London: Rowman & Littlefield), pp. 51–67.

Palmer, C. 2010. *Animal Ethics in Context* (New York: Columbia University Press).

Palmer, C. 2015. "Against the View That We Are Normally Required to Assist Wild Animals." *Relations* 3: 203–10.

Palmer, C. 2019 (online first). "Assisting Wild Animals Vulnerable to Climate Change: Why Ethical Strategies Diverge." *Journal of Applied Philosophy*.

Parfit, D. 1984. *Reasons and Persons* (Oxford: Oxford University Press).

Pearce, D. 2015. "Reprogramming Predators." Available at: www.hedweb.com/abolitionist-project/reprogramming-predators.html (Accessed on Jan 1st, 2019).

Pepper, A. 2019. "Adapting to Climate Change: What We Owe to Other Animals." *Journal of Applied Philosophy* 36: 592–607.
Regan, T. 2004. *The Case for Animal Rights* (Berkeley: University of California Press).
Sebo, J. "Animals and Climate Change." Forthcoming in *Philosophy and Climate Change*. Eds. M. Budolfson, T. McPherson, and D. Plunkett (Oxford: Oxford University Press).
Shiffrin, S. 1999. "Wrongful Life, Procreative Responsibility, and the Significance of Harm." *Legal Theory* 5: 117–48.
Simmons, A. 2009. "Animals, Predators, the Right to Life, and the Duty to Save Lives." *Ethics and the Environment* 14: 15–27.
Singer, P. 1975. *Animal Liberation* (New York: Avon Books).

# 4

# ASSESSING THE RISKS OF INTERVENTION

## 4.1 Introduction

In light of their growing appreciation for the extent of wild animal suffering (WAS), a number of animal rights theorists, myself included, have adopted a qualified commitment to humanitarian intervention in the wild. More specifically, we've adopted a view I call 'fallibility-constrained interventionism.'[1] According to this view, preventing WAS is desirable and thus, we should intervene in nature, but we should proceed very cautiously in light of our limited understanding of ecosystems and the resulting ecological risks intervention poses.[2] For example, in previous work, I argue that a relatively new form of gene editing called CRISPR presents a promising means of intervention.[3] I claim that CRISPR could hypothetically be used to modify r-strategists so that they become more like K-strategists, that is, so that they produce a fewer number of offspring and invest the energy needed to ensure that their offspring have a decent shot at life. What's more, I note that CRISPR-created traits like the above could easily be dispersed through wild animal populations via an 'endonuclease gene drive.'[4] However, I don't think that we should immediately begin to bioengineer wildlife populations. A considerable amount of ecological research, as well as biotechnological research and testing, is required before any large scale genetic intervention would be ecologically safe. It's also likely that supplementary interventions would have to be researched in order to avoid ecological catastrophe, for example, engineering an alternative food source for predators who, due to the initial intervention, no longer have as many r-strategists to eat.

In this chapter,[5] I explain and respond to two objections to fallibility-constrained interventionism. Both objections attack the claim that cautious intervention is justified, but they do so on opposite fronts. According to the first objection, cautious intervention is unjustified because fallibility is allegedly intractable. Ecosystems' complexity entails not only that large-scale interventions are at risk of causing unintended negative consequences, but that we're unable to reliably assess the extent of that risk.[6] If this is right, then it would seem that proceeding both cautiously and on a large scale is impossible. Even the most well-researched, large-scale intervention could be ecologically dangerous and thus, only very small-scale interventions in nature are ever justified, for example, rescuing deer trapped in barbed wire fences or helping beached sea mammals return to the ocean.

In contrast, the second objection claims that adopting a cautious attitude reflects a failure to appreciate the moral implications of WAS. From what we can tell about wild animal reproduction and the conditions wild animals face, most wild animals experience more suffering than enjoyment in their lives, that is, their lives contain net suffering.[7] If this is right, it seems odd to worry about the possibility that large-scale interventions could unintentionally damage ecosystems. In fact, ecological damage could have positive consequences. Destroying wild animals' habitats would reduce the number of wild animals in the world, and since most wild animals' lives contain net suffering, reducing the number of wild animals would increase total utility. The upshot, it would seem, is that instead of carefully trying to avoid ecological damage, we should intentionally destroy wild animals' habitats.[8]

In this chapter, I grant, for the sake of argument, both the claim that fallibility is intractable and the claim that (some forms of) habitat destruction have positive consequences overall. I argue that even if one grants these claims, they fail to undermine a cautious commitment to large-scale intervention. With respect to the second claim, even if nature (or the world) contains moral disvalue, intentional habitat destruction is only justified from a utilitarian perspective. On the plausible, non-utilitarian assumption that negative duties are much weightier than positive duties, our duty to refrain from causing the harms associated with habitat destruction trumps our duty to provide the benefits associated with it. However, that habitat destruction could have positive consequences overall, is morally significant. Any possible positive effects of ecological damage should be considered when morally evaluating the ecological risks of large-scale intervention. Furthermore, the claim that fallibility is intractable suggests that any unintended harms caused by large-scale intervention are not only unintended, but unforeseeable. From a deontological perspective, unintended, unforeseeable harm is more excusable than other kinds of harm. In combination, epistemic intractability and the positive consequences produced by some forms of habitat destruction suggest that we should proceed cautiously but not be paralyzed by our limited understanding of ecosystems.

## 4.2 Intentional Habitat Destruction

I'll begin with the second objection first. According to it, the extent of WAS is too severe for caution to make sense. Wild animals aren't just experiencing *a lot of suffering*: in most cases, they are experiencing *net suffering*, that is, their lives contain a greater amount of suffering than enjoyment.[9] The upshot is that most (sentient) wild animals are better off never having been born at all. According to Brian Tomasik, this observation implies that the sort of intervention we should be pursuing is habitat destruction.[10] Destroying wild animals' habitats would reduce the number of wild animals who are born, thereby reducing the amount of moral disvalue in nature. What's more, the longer we wait the worse things will be, since waiting just gives nature the opportunity to produce more generations of doomed animals.

A number of issues are worth mentioning with respect to Tomasik's argument. The first is whether he's right to claim that most wild animals don't live lives worth living. Given the prevalence of the r-strategy and the manner in which most r-strategist young die, it's highly probable that most wild animals live pretty terrible lives. However, a pretty terrible life isn't the same thing as a life not worth living.[11] As noted in Chapter 2, people who are in absolute poverty face pretty terrible conditions, but it's still likely that most of them live lives worth living, as most have the opportunity to form friendships, develop talents, etc. Of course, most r-strategists don't get to enjoy rewarding experiences. They just crawl around uncomfortably for a bit and then die painfully. Still, it's somewhat unclear whether their lives are merely pretty terrible (certainly not flourishing lives), or whether their lives aren't worth living at all.

For the sake of argument, let's assume that Tomasik is right to claim that most wild animals don't live lives worth living. The implications of this claim for overall utility remain an issue. If most wild animals experience more suffering than enjoyment in their lives, does that mean that the total amount of utility in nature is negative? That nature contains net suffering? Tomasik and others seem to think so. Furthermore, does it mean that the overall amount of utility in the world is negative, that is, that the overall amount, considering the utility of human beings and domesticated animals as well, is negative?[12] On the one hand, it's possible that the comparatively enjoyable lives that most human beings and *some* domesticated and wild animals live, outweigh the miserable lives that most r-strategists live. However, on the assumption that most r-strategist young don't live lives worth living, I think it's unlikely that their misery is outweighed. For one, some philosophers, including Tomasik,[13] maintain that there's an asymmetry between suffering and enjoyment. Though some extreme versions of this view deny that enjoyment has any positive value at all, the more moderate versions merely maintain – quite reasonably, I think – that the badness of an amount of suffering exceeds the goodness of an equal (or only marginally greater) amount of enjoyment.[14] If, when calculating utility, a unit of suffering is accorded more negative value than a unit of enjoyment is positive value, it's quite likely that the suffering of r-strategists outweighs the enjoyment of K-strategists. But even if we deny that there's an asymmetry between suffering and enjoyment, it's unlikely that the suffering of r-strategists is outweighed. After all, the number of r-strategist individuals in the world dwarfs the number of K-strategist individuals (human beings included). This remains true even if we restrict our count to r-strategist mammals, and it is especially true if we include in our count the number of individuals who were alive at some point during a period of years – a few decades let's say – instead of the number of individuals alive at any particular instant. Though there's still some room for doubt, the claim that most r-strategist young don't live lives worth living plausibly entails that the overall amount of utility in the world is negative. And if most r-strategists don't live lives worth living, then reducing their populations via habitat destruction promises to reduce net suffering.

There are, of course, many harms associated with habitat destruction. The manner in which habitat destruction prevents disvaluable lives from coming into existence is specifically by killing existing wild animals. The idea is that without their habitats and the resources habitats contain, existing r-strategist animals will die and will therefore be prevented from reproducing. Furthermore, habitat destruction would foreseeably cause considerable harm to K-strategist animals (deer, bears, elephants, etc.) and to human beings (we rely on wild animal habitats in various ways, some of which I take note of below). However, if the r-strategy does indeed entail that the overall amount of utility in the word is negative, it's likely that the benefits of a significant, enduring reduction in the size of r-strategist populations exceed any harms to existing (or future) individuals.

One might object that the case for habitat destruction presupposes that our goal should be to increase total (aggregate) utility, rather than average utility. After all, reducing the size of r-strategist populations only increases average utility when doing so reduces the ratio of r-strategists to K-strategists. Since habitat destruction reduces the size of K-strategist populations, too, it's likely that, in many cases, it would leave the ratio of r-strategists to K-strategists more or less intact. In such cases, the immediate harms habitat destruction cause would not be outweighed by greater, compensating benefits, at least from a perspective focused on average utility.

The view that our goal should be to promote *average* utility, rather than *total* utility, is quite plausible when dealing with individuals whose net utility level is positive, that is, individuals whose lives are worth living. This is because making total utility our goal counterintuitively implies that it would be good to bring into existence as many individuals as possible, so long as they live minimally decent lives;[15] or in other words, that we should prefer large, unhappy populations (where total utility is high but average utility is low) over small, happy populations (where average utility is high but total utility is low). However, when dealing with individuals whose lives contain more suffering than enjoyment, considerations of total utility have special importance. To see why, compare two possible worlds with the same average level of net suffering, but one of which has a much higher population than the other. More specifically, let's say that the first possible world contains five people, all of whose lives are not worth living, whereas the second possible world contains 10,000 people, all of whose lives are not worth living. We should all be able to agree that the world with five people in it is preferable to the world with 10,000 people in it. However, the reason can't be differential average utility, since we've stipulated that each possible world possesses the same level of it. Instead, the reason must be differential total utility: the world with a population of 10,000 has a much higher total level of net suffering, and that's why it's worse than the world with five people in it. I'd even go so far as to say that consideration of total utility can *trump* average utility, at least when we're dealing with net suffering. To see why, compare two possible worlds with different average levels of net suffering, one of which has a much higher population, but also a lower average level of net suffering. More specifically, let's say that the first possible world contains five people, all of whose lives are not worth living, and that the second possible world contains 10,020 people, 10,000 of whose lives are not worth living but 20 of whose lives aren't too bad – a world not unlike the one we actually live in, if r-strategists lives aren't worth living. Even though the second possible world has a lower average level of net suffering, it's still a worse world than the first, because the second world's total level of net suffering is much higher. In light of this, it's appropriate for our goal to be the reduction of total, rather than average net suffering. If I'm right, then Tomasik's case for habitat destruction can't be deflected so easily since habitat destruction promises to reduce total net suffering by reducing the size of r-strategist populations.

It should be noted that only some kinds of habitat destruction consistently reduce the size of r-strategist populations. Habitats that are significantly disrupted can become inhospitable to some species, particularly specialist species, without becoming inhospitable to other species. As a result, habitat destruction can sometimes lead to population increases among generalist species, including generalist r-strategists.[16] In order for habitat destruction to reliably decrease the size of r-strategist populations, then, it has to be the sort of destruction that renders a region thoroughly inhospitable for the majority of wild animals, for example, desertification. It's specifically habitat destruction of this sort: the sort that dramatically reduces a region's biomass and primary productivity, that Tomasik has in mind.

For some readers, the claim that we should intentionally destroy habitat will seem absurd regardless of whether doing so improves overall utility. After all, the 'beneficiaries' of habitat destruction are the wild animals we thereby prevent from being born, not human beings. In fact, habitat destruction is clearly not in most human beings' interests. We benefit in various ways from many wild animal habitats, for example, wild animal habitats are often aesthetically pleasing, they often contain economically valuable resources, the plants they contain produce oxygen that we enjoy breathing, etc. Thus, it might be argued that habitat destruction is unjustified because it's contrary to the interests of human beings. Of course, this counter-argument assumes human exceptionalism, that is, it assumes that the interests of human beings matter more than the interests of animals. I think it's more interesting to ask whether Tomasik's conclusion can be plausibly rejected without assuming human exceptionalism. I think this, in part, because Tomasik is not a human exceptionalist and neither is much of his audience. I also think this because I think human exceptionalism is false.

The most obvious, non-human-exceptionalist reply to Tomasik's argument is that it assumes utilitarianism, and utilitarianism is not very plausible. Whether our actions produce positive consequences (or reduce disvalue) matters, of course, but it's not the only thing that matters. In particular, negative duties are weightier than positive duties and, as a result, doing the right thing sometimes requires that we refrain from performing the action that produces the best consequences. For example, in a situation where killing one person is necessary in order to save two people, killing that person would produce the best consequences. But that doesn't mean saving two people justifies killing one. Your positive duty to save two people is trumped by your negative duty to refrain from killing one. My own view is that, past some threshold, positive duties do trump negative duties, for example, that saving 10 billion people would justify killing one.[17] Whatever the threshold is, though, it's high enough that the duty to refrain from killing typically (but not always) trumps the duty to save lives, and I think that what's typically the case holds true with respect to habitat destruction. In fact, following John Hadley, it's plausible to think that wild animals have something like a property right to their habitat. As he points out, wild animals require access to their habitat to satisfy their basic needs, and since we think that our own basic needs (as well as some of our less significant interests) are sufficient to ground property rights in the human case, it seems speciesist to deny that wild animals have a right to their habitat.[18]

Tomasik thinks that the positive consequences of habitat destruction exceed the harms it would cause to existing animals and human beings, and he may be right about that. But on the plausible assumption that negative duties are more stringent than positive duties, it's unlikely that our duty to provide the benefits associated with habitat destruction trumps our duty to respect the interests of existing sentient beings. That Tomasik thinks otherwise reflects his commitment to utilitarianism.

Though I disagree with the conclusion that we ought to intentionally destroy habitat, I agree that, on the assumption that nature contains net suffering, the sort of habitat destruction he has in mind likely has positive consequences overall. Furthermore, the possibility that some forms of habitat destruction have positive consequences overall, has implications for the second objection to fallibility-constrained interventionism, that is, the objection that large-scale interventions in nature are reckless even if carefully conducted.

## 4.3 Intractable Fallibility

The second objection to fallibility-constrained interventionism was put forward by Nicolas Delon and Duncan Purves.[19] They claim that those who see fallibility as a constraint to intervention, rather than as a straightforward bar to intervention, are underestimating the extent of our fallibility. After all, in order for fallibility to constrain but not entirely prevent intervention, it must currently, or at least eventually, be possible for us reliably to make *some* judgments concerning the effects of intervention. It's probably too much to hope that we'll ever be able to know precisely what the ecological effects of a given large-scale intervention would be, but perhaps we can write up a list of possible effects and, with enough research, we might be able to reliably assign probabilities to them, along with error margins. If we can bring ourselves to the point where we can reliably predict the possible effects, their probabilities, and the error margins associated with those probabilities, then fallibility would be quite manageable. Even if we'll never be able to say precisely what's going to happen when we implement an intervention, we would be able to distinguish between interventions with acceptable risks and interventions with unacceptable risks.[20]

According to Delon and Purves, however, there's good reason to think that fallibility isn't and may never become epistemically tractable. Their claim is partially justified in light of recent work in contemporary ecology that suggests ecosystems are more complex and less predictable than ecologists used to think. The main idea is that significant changes to an ecosystem can reduce its resilience, that is, its ability to absorb future disruptions without shifting to a new regime. What this means is that assessing the risks of an intervention involves more than just determining its possible direct effects: we also have to determine its possible indirect effects. More specifically, we would have to determine how likely it is that, post-intervention, an ecosystem will be able to handle future disruptions to which we may have rendered it more vulnerable; and we'd have to determine and compare the effects that different possible regimes could have on animal and human welfare: a difficult set of epistemic tasks.[21]

Additionally, they note that climate change undermines our ability to make reliable assessments.[22] Our assessment of the possible ecological effects of intervention depend upon our knowledge of a given ecosystem's current state, but climate change threatens to affect many ecosystems significantly. We could, of course, make assessments that assume climate change, but it's difficult to know what the future climate situation will be, as that situation depends on a number of variables concerning collective human behavior. Perhaps we'll change our collective behavior and emissions will drop considerably over the next 20 years. Or perhaps we won't. That and other climate change related matters are difficult to predict with much accuracy.

As with the previous objection, I'm going to grant Delon and Purves's worry for the sake of argument. My question is the following: even if it's true that we'll never be able to reliably list all of the possible direct and indirect effects that an intervention could have, their probabilities, or the error margins associated with those probabilities, does it follow that we ought never to intervene on a large scale?[23] I don't think it does follow, and I have two main reasons.

First, we noted in the previous section that the consequences of ecological damage are complicated. Admittedly, habitat destruction always causes some harm to existing wild animals, but in cases where it causes a region to become generally inhospitable, it also has the effect of reducing the size of r-strategist populations and thus (probably), increasing total utility. In cases where a habitat is merely disrupted, however, r-strategist populations may not be reduced and could actually be increased, in so far as r-strategists also tend to be adaptable generalists. Unless we can tell in advance how the damage being risked will affect r-strategist populations, it isn't possible to say whether that damage, if it eventuates, will produce an overall good or an overall bad outcome. That there's uncertainty about how we should morally evaluate the outcomes being risked is something that should be taken into consideration when determining how best to respond to those risks. The possibility that unintended ecological damage would, if it eventuated, have overall positive consequences, lends justificatory support to the claim that fallibility should not paralyze efforts to develop and eventually implement large-scale interventions.

Of course, it may turn out to be the case that most r-strategists merely live terrible lives instead of lives that contain more suffering than enjoyment. If it turns out that most r-strategists merely live terrible lives, then it's no longer likely that nature contains net suffering. And if nature doesn't contain net suffering, then habitat destruction (of the sort that reduces r-strategist populations) won't have positive consequences overall. Though the case for large-scale intervention is somewhat stronger if nature contains net suffering, the argument is defensible even if nature doesn't. For one, the suffering of r-strategists would still be a strong reason to intervene. Successfully using gene drives to reduce r-strategist populations and increase parental care would significantly benefit r-strategist young, thereby increasing the level of average utility in nature. Of course, gene drives, even when conducted carefully, could end up causing unintended ecological damage. But unintended damage is excusable under the right conditions, or so I argue below.

My second reason for rejecting Delon and Purves' conclusion concerns the harms associated with ecological damage. As mentioned earlier, Delon and Purves think that an inability to make reliable risk assessments implies that we should refrain from large-scale intervention. It should be noted, however, that an inability to anticipate the possible negative effects of intervention has implications for how those effects should be categorized. Any harmful side effects caused by intervening in the wild would not only be unintentional, but unforeseeable as well. This is important because whether a harm is intentional, and whether a harm is foreseeable, both matter for our moral assessment of it. Intentional harms are obviously the worst kind, for example, someone who intentionally strikes or kills another person is guilty of assault or murder. Unintended harms, by contrast, are more easily excused. But whether an unintended harm is excusable depends on whether it was foreseeable. For example, a doctor who accidentally kills a patient is guilty of malpractice if the effects of their treatment were foreseeable. Perhaps the patient's medical records indicate that he/she's allergic to the medication administered and thus, his/her doctor should have been aware of his/her allergy. Conversely, if the doctor had no way of knowing that their patient was allergic, we would normally say that the doctor isn't at fault.

The manner in which we should assess any unintended, negative consequences caused by intervention in nature is roughly analogous to the manner in which we assess the side effects of medical treatment. Like medical treatment, intervention in nature is meant to be beneficial. And like medical treatment, it can produce unintended negative consequences. Since we normally excuse cautious medical practitioners when they accidentally cause harms that could not reasonably have been foreseen, we should have the same attitude with respect to intervention in nature. Of course, there's a question about what qualifies as a sufficient degree of caution. For example, we might wonder how much time and resources should be devoted to testing and to ecological research before a large-scale intervention can be responsibly conducted. Though it may be tempting to think that centuries of testing and research are needed,[24] I think that a somewhat less conservative attitude is warranted. Consider again the analogy with medical treatment. Though we expect doctors to be cautious when deciding upon and administering a treatment, we also think that the appropriate degree of caution is sensitive to how urgent the situation is. In a medical emergency where the patient will soon die without intervention, we'd normally say that fast action on the doctor's part is morally required. But if that's the conclusion we draw in the medical case, we should draw a similar conclusion about intervention in the wild because waiting to intervene has significant negative consequences. Waiting means that more generations of r-strategists will come into existence and live terrible lives, generations that we might have prevented from being born had we acted more quickly. Though we should wait until we've developed a degree of competence before intervening (much the way we expect those who respond to medical emergencies to undergo medical training), we needn't wait until we've achieved perfect reliability. We may never be able to make perfectly reliable judgments about the ecological risks of intervening in nature, and even if it is possible to achieve perfect reliability after centuries of testing and research, hundreds of years is too long to wait. Nature has created a perpetual, humanitarian emergency, and we should aim to intervene within our lifetime.

## 4.4 Conclusion

In summary, this chapter has responded to two objections to fallibility-constrained interventionism. According to the first objection, taking a cautious attitude toward large-scale intervention is inconsistent with fully appreciating the extent and moral significance of WAS. The objection states that we should embrace ecological destruction, rather than avoid it, since destroying habitats would reduce the size of wild animal populations, thereby reducing the amount of net suffering in nature.

In contrast, the second objection states that fallibility isn't tractable enough to function as a constraint. Ecosystems are too unpredictable for us to make reliable judgments about the risks of large-scale intervention, so instead of constraining large-scale intervention, fallibility should bar it.

In reply, I've argued that the first objection, though flawed, contains an important truth that helps us to address the second objection. More specifically, intentional ecological destruction is unjustified because our duty to refrain from causing the harms associated with it trumps our duty to provide the benefits. I acknowledge, however, that ecological destruction could potentially produce good consequences overall (by reducing r-strategist populations), and that this should be considered when morally evaluating the risks of intervention. Furthermore, unlike the harms associated with intentional destruction, the harms associated with unintentional destruction are excusable, so long as the destruction was also unforeseeable. And that's exactly what epistemic intractability implies: that even with considerable testing and research, the ecological risks of large-scale intervention are, to some extent, unforeseeable.

In light of the above, the correct conclusion is that we should maintain a cautious commitment to humanitarian intervention in nature. Large-scale interventions such as CRISPR gene drives promise to benefit wild animal populations significantly, and though the ecological risks of intervention are difficult to assess accurately, our fallibility shouldn't be paralyzing. The consequences of ecological destruction could potentially be positive overall, and insofar as ecological destruction is both unintentional and unforeseeable, then the harms associated with it are also excusable.

## Notes

1 Johannsen (2017) p. 333.
2 See Cowen (2003), McMahan (2010) and (2015 ), Sözmen (2013), Horta (2013), (2015), and (2017), Tomasik (2015b), and Johannsen (2017).
3 Johannsen (2017) pp. 339–43. The acronym 'CRISPR' stands for 'Clustered Regularly Interspaced Short Palindromic Repeats.' Though the term refers to a structural feature present in the genomes of different bacteria, it's now colloquially used to refer to a new form of gene editing (Doudna and Charpentier, 2014 p. 1 and p. 3). For an informative and accessible article about CRISPR, see Ledford (2015).
4 For descriptions of how gene drives work, see Ledford (2015) p. 22 and Esvelt et al. (2014) pp. 3–4. For examples of successful gene drives that have been conducted in the laboratory setting, see Gantz et al. (2015) and Hammond et al. (2016). For some speculative thoughts about using gene drives to aid wild animals, see Esvelt (2018).
5 This chapter is basically a shortened version of Johannsen (2020).
6 Delon and Purves (2018) pp. 244–50. Though the authors I cite in endnote 2 understand fallibility specifically as a constraint on intervention, the issue of fallibility is often raised in objection to the claim that intervention is justified. See, for example, Singer (1975) pp. 238–9, Simmons (2009), and Ladwig (2015) pp. 297–9.
7 For work supporting the claim that most wild animals experience net suffering, see Ng (1995), Horta (2010) and (2015), and Tomasik (2015b).

8 Tomasik (2016).
9 See endnote 6.
10 Tomasik (2016). See also his somewhat more reserved comments about habitat destruction in Tomasik (2015b) pp. 145–6.
11 I assume that a life with more suffering than enjoyment is not a life worth living. However, I don't think that experiencing a lot of pleasure is sufficient to live a flourishing life, i.e., I don't have an entirely hedonic view of the good. A certain amount of pleasure is a necessary condition for flourishing, but the good is probably a plural thing containing a number of objective elements. For an account of (specifically human) flourishing that strikes me as plausible, see Sypnowich (2017) chapter 7.
12 I might also have included liminal animals in this list. Liminal animals are undomesticated animals that live in human communities. See Donaldson and Kymlicka (2011) chapter 7.
13 Tomasik (2015a).
14 See, for example, Mayerfeld (1996) pp. 324–5.
15 There's a good discussion of this in Rawls (1971) pp. 161–4.
16 For a relevant study that focuses specifically on bird populations, see Devictor et al. (2008).
17 This view is sometimes labelled 'Moderate Deontology.' For examples of this view, see Ross (1930) pp. 88–9 and Nagel (1991) pp. 62–3.
18 Hadley (2015) pp. 53–8. For a Lockean account of wild animals' right to habitat, see Milburn (2017).
19 Delon and Purves (2018) pp. 244–50.
20 Another way of putting it is that we would have much of the information necessary for cost-benefit analysis, i.e., information necessary to calculate the expected value of a proposed intervention. For an introduction to cost-benefit analysis and other important ideas in decision theory, see, for example, Resnik (1987).
21 Delon and Purves (2018) pp. 244–5. See also Walker and Salt (2006).
22 Delon and Purves (2018) pp. 248–50.
23 The principle that we should avoid performing actions whose risks are uncertain, is normally referred to as the precautionary principle. For a discussion of this principle, see Steel (2015).
24 Delon and Purves speculatively concede that, perhaps after centuries of research, the epistemic difficulties associated with large-scale intervention could be overcome. In particular, they suggest that devoting research to 'robustness analysis' – a method used to determine which theoretical models make reliable predictions – could be helpful. See Delon and Purves (2018) pp. 256–7 and Weisberg (2006).

## References


Cowen, T. 2003. "Policing Nature." *Environmental Ethics* 25: 169–82.

Delon, N., and D. Purves. 2018. "Wild Animal Suffering is Intractable." *Journal of Agricultural and Environmental Ethics* 31: 239–60.

Devictor, V., R. Julliard, and F. Jiguet. 2008. "Distribution of Specialist and Generalist Species Along Spatial Gradients of Habitat Disturbance and Fragmentation." *Oikos* 117: 507–14.

Donaldson, S., and W. Kymlicka. 2011. *Zoopolis: A Political Theory of Animal Rights* (New York: Oxford University Press).

Doudna, J.A., and E. Charpentier. 2014. "The New Frontier of Genome Editing with CRISPR-Cas9." *Science* 346: 1258096.

Esvelt, K.M., A.L. Smidler, F. Catteruccia, and G.M. Church. 2014. "Concerning RNA-Guided Gene Drives for the Alteration of Wild Populations." *eLife* 3: e03401.

Esvelt, K.M. 2018. "When Are We Obligated to Edit Wild Creatures?" *Leapsmag* 28 May 2018. Available at: www.leapsmag.com/when-are-we-obligated-to-edit-wild-creatures/ (Accessed on 5th April, 2019).

Gantz, V.M., N. Jasinskiene, O. Tatarenkova, A. Fazekas, V.M. Macias, E. Bier, and A.A. James. 2015. "Highly Efficient Cas9-Mediated Gene Drive for Population Modification of the Malaria Vector Mosquito *Anopheles Stephensi*." *PNAS* 112: E6736–E6743.

Hadley, J. 2015. *Animal Property Rights: A Theory of Habitat Rights for Wild Animals* (Lanham: Lexington Books).

Hammond, A., R. Galizi, K. Kyrou, A. Simoni, C. Siniscalchi, D. Katsanos, M. Gribble, D. Baker, and E. Marois. 2016. "A CRISPR-Cas9 Gene Drive System Targeting Female Reproduction in the Malaria Mosquito Vector *Anopheles Gambiae*." *Nature Biotechnology* 34: 78–83.

Horta, O. 2010. "Debunking the Idyllic View of Natural Processes." *Télos* 17: 73–88.

Horta, O. 2013. "*Zoopolis*, Intervention, and the State of Nature." *Law, Ethics, and Philosophy* 1: 113–25.

Horta, O. 2015. "The Problem of Evil in Nature: Evolutionary Bases of the Prevalence of Disvalue." *Relations* 3: 17–32.

Horta, O. 2017. "Animal Suffering in Nature: The Case for Intervention." *Environmental Ethics* 39: 261–79.

Johannsen, K. 2017. "Animal Rights and the Problem of r-Strategists." *Ethical Theory and Moral Practice* 20: 333–45.

Johannsen, K. 2020. "To Assist or Not to Assist? Assessing the Potential Moral Costs of Humanitarian Intervention in Nature." *Environmental Values* 29: 29–45.

Ladwig, B. 2015. "Against Wild Animal Sovereignty: An Interest-Based Critique of *Zoopolis*." *The Journal of Political Philosophy* 23: 282–301.

Ledford, H. 2015. "CRISPR, the Disruptor." *Nature* 522: 20–4.

Mayerfeld, J. 1996. "The Moral Asymmetry of Happiness and Suffering." *The Southern Journal of Philosophy* 34: 317–38.

McMahan, J. 2010. "The Meat Eaters." *New York Times Opinionator* 19 September 2010. Available at: www.opinionator.blogs.nytimes.com/2010/09/19/the-meat-eaters/ (Accessed on 5th April, 2019).

McMahan, J. 2015. "The Moral Problem of Predation." In *Philosophy Comes to Dinner: Arguments about the Ethics of Eating*. Eds. A. Chignell, T. Cuneo, and M. Halteman (London: Routledge), pp. 268–94.

Milburn, J. 2017. "Nonhuman Animals as Property Holders: An Exploration of the Lockean Labour-Mixing Account." *Environmental Values* 26: 629–48.

Nagel, T. 1991. "War and Massacre." In *Mortal Questions*. By T. Nagel (New York: Cambridge University Press), pp. 53–74.

Ng, Y.K. 1995. "Towards Welfare Biology: Evolutionary Economics of Animal Consciousness and Suffering." *Biology and Philosophy* 10: 255–85.

Rawls, J. 1971. *A Theory of Justice* (Cambridge, MA: Belknap Press).

Resnik, M. 1987. *Choices: An Introduction to Decision Theory* (Minneapolis: University of Minnesota Press).

Ross, W.D. 1930. "What Makes Right Acts Right?" In *The Right and the Good* (Oxford: Oxford University Press). Reprinted in *20th Century Ethical Theory*. Eds. S.M. Cahn and J.G. Haber (Upper Saddle River: Prentice-Hall, 1995), pp. 87–105.

Simmons, A. 2009. "Animals, Predators, the Right to Life, and the Duty to Save Lives." *Ethics and the Environment* 14: 15–27.

Singer, P. 1975. *Animal Liberation* (New York: Avon Books).

Sözmen, B.İ. 2013. "Harm in the Wild: Facing Non-Human Suffering in Nature." *Ethical Theory and Moral Practice* 16: 1075–88.

Steel, D. 2015. *Philosophy and the Precautionary Principle* (Cambridge: Cambridge University Press).

Sypnowich, C. 2017. *Equality Renewed: Justice, Flourishing, and the Egalitarian Ideal* (New York: Routledge).

Tomasik, B. 2015a. "Are Happiness and Suffering Symmetric?" *Essays on Reducing Suffering*. Available at: www.reducing-suffering.org/happiness-suffering-symmetric/#Small_is_beautiful (Accessed on 5th April, 2019)

Tomasik, B. 2015b. "The Importance of Wild Animal Suffering." *Relations* 3: 133–52.

Tomasik, B. 2016. "Habitat Loss, not Preservation, Generally Reduces Wild Animal Suffering." *Essays on Reducing Suffering*. Available at: www.reducing-suffering.org/habitat-loss-not-preservation-generally-reduces-wild-animal-suffering (Accessed on 5th April, 2019).

Walker, B., and D.A. Salt. 2006. *Resilience Thinking: Sustaining Ecosystems and People in a Changing World* (Washington, D.C.: Island Press).

Weisberg, M. 2006. "Robustness Analysis." *Philosophy of Science* 73: 730–42.

# 5

# EDITING NATURE

## 5.1 Introduction

In previous chapters, I hope to have established that there's a strong case for large-scale, humanitarian intervention in nature. Predation, the r-strategy, and other natural processes are responsible for a tremendous amount of suffering, and though the risks of intervention must certainly be considered, they don't suffice to undermine a cautious commitment to large-scale assistance.

In this chapter, I'll argue that genetically editing wild animal populations is a type of intervention especially worthy of research. Particular attention will be devoted to CRISPR,[1] a recent type of gene editing with the power to significantly change the natural world. In Section 5.2, I describe how CRISPR works and explain why it's such a powerful form of intervention. Though CRISPR has various virtues, its greatest one, arguably, is that CRISPR-created traits can be designed so that they're inherited by nearly all of an edited organism's descendants. In other words, CRISPR can 'drive' traits through wild populations.[2]

Section 5.3 compares conventional wildlife management to gene drives. In it, I argue that the moral costs of the former are far greater than those of the latter.[3] I also respond to the objection that gene drives, even if better than conventional management, are impermissible because developing them requires experimenting on animals who are capable of suffering. I argue that such experiments are justified so long as we're also justified in believing that they'll lead to the implementation of successful, large-scale interventions, and I recommend that initial experiments be conducted on animals who lack the capacity to suffer.

Section 5.4 compares a number of different goals that gene drives could potentially be used to try to achieve: (a) to make certain species extinct;[4] (b) to remove the capacity to suffer from certain animals;[5] or (c) to change animals' dietary and reproductive behaviors.[6] Though I argue that behavior change is ideal, I concede that it's sensible to eliminate at least some species of non-sentient parasite. Furthermore, I concede that, in the event that it proves infeasible to safely change wild animals' dietary and reproductive behaviors, removing certain animals' capacity to suffer is a second-best option worth exploring.

## 5.2 CRISPR

One especially promising type of intervention is gene editing. CRISPR, the latest form of gene editing, has received a considerable amount of attention in the media over the last four years, and for good reason. For one, it's both cheap and easy to use. Whereas earlier forms of gene editing required costly materials and a considerable amount of specialized expertise in protein engineering, CRISPR has comparably low material costs, and it edits the desired part of a genome via an easily modified RNA molecule.[7] This comparative inexpensiveness and ease of use has democratized gene editing, making it possible for labs of modest means to use gene editing in their research. In fact, CRISPR is so accessible that amateurs called 'biohackers' are able to use it,[8] a fact that was dramatically demonstrated when Josiah Zayner live-streamed himself self-administering a CRISPR-based muscle enhancer. Since Zayner's live-stream, a number of other live-streamed publicity stunts have been pulled by biohackers. Though Zayner himself has decided to abstain from further stunts, he still involves himself in a variety of CRISPR-related activities.[9] Among other things, he owns a company called 'The Odin' that sells do-it-yourself CRISPR kits. The kits available for purchase include one for creating fluorescent yeast, and one that can be used to create 'bioart' via different colored bacteria.[10]

A second reason for all the media attention is that CRISPR-created traits can be quickly dispersed through a target population via a 'gene drive.' In general, the term 'gene drive' refers to the process that occurs when a gene increases its odds of being passed on. There are a number of methods via which genes may do this,[11] but the method associated with CRISPR is called an 'endonuclease drive.' Endonuclease drives occur when a gene possessed by one of a pair of chromosomes cuts the associated area of the partner chromosome. The cell, in turn, repairs the damage to the partner chromosome by copying the 'attacking' gene onto it. The gene's subsequent presence in both chromosomes ensures that it will be inherited by nearly all of its organism's offspring.[12] The reason CRISPR is well-suited for implementing endonuclease drives is because it works in a similar manner: it cuts the targeted part of the genome, and this cut is in turn repaired using an edited gene. Via an endonuclease gene drive, then, CRISPR can spread a genetically engineered trait through a target population, and a number of studies have already done so with insects, albeit in a laboratory setting.[13]

In combination, CRISPR's inexpensiveness and ability to start gene drives make it a potentially powerful tool for assisting wild animals. Traits that would benefit wild animals could be developed in the laboratory and then spread through wild animal populations. Though it's hard to predict exactly which traits we'll be able to develop, CRISPR's record of success in modifying insects and animals gives us reason to be optimistic. Consider the research being done on mosquitos. It has already been demonstrated, in the laboratory setting, that modifications which spread sterility through female mosquitos (thereby reducing population size) or which make mosquitos resistant to malaria parasites, can be driven through mosquito populations.[14] And the animal farming industry has had some success at modifying cows in ways intended to make both raising and killing them more 'humane.' Examples include the development of cows who are resistant to bovine tuberculosis,[15] and the development of bulls that don't grow horns and thus, who won't eventually be made to undergo a painful dehorning.[16] Whether welfarist modifications are morally good, all things considered, is an open question, but they certainly demonstrate CRISPR's huge potential for producing new phenotypes.[17]

## 5.3 Justifying the Moral Costs of Gene Drives

Gene drives are appealing not only in light of their potential for benefiting wild animals, but also because the moral costs of using them are small relative to conventional wildlife management. When writers imagine what it would mean to assist wild animals on a large scale, they often imagine a zoo-like scenario that involves significant restrictions on wild animals' liberties.[18] Examples of such restrictions include using fences to separate predators from their prey, and feeding predators a plant-based source of protein; as well as using sterilization or contraceptives to restrict the reproduction of all but a small number of r-strategist individuals and subsequently caring for those infants who are born.

The main reason why gene drives are preferable to conventional wildlife management is because the harms of gene drives are initial, rather than perpetual, and because those harms are inflicted on only a small subset of each target population, rather than the entire population. As noted earlier, gene drives are conducted by modifying a small number of edited organisms who, upon release into the wild, spread the relevant trait or traits throughout wild populations. Though being experimented upon and held captive in a lab are certainly significant harms for the animals who must undergo it, only a small proportion of the current population are required to endure these harms, and future members of the population needn't endure any harms at all. In comparison with a zoo-like scenario, then, the moral costs associated with gene drives are quite small. Using gene drives to assist wild animals is compatible with leaving most wild animals free to live their lives.

It might be objected that gene drives, though less morally costly than conventional wildlife management, are still too costly to be permissible. One obvious worry is the possibility of unintentionally causing significant ecological damage. I discussed fallibility at length in Chapter 4, so I won't discuss it much here. However, I will say that significant research and testing would have to be conducted before beneficent gene drives could be conducted responsibly. I'll also note that research has already been conducted on 'daisy-chain' gene drives – a technique that promises to allow for localized drives that don't spread beyond a certain geographical area.[19] Unlike a standard gene drive, a daisy-chain drive operates via a series of linked elements. Each element in the chain drives the next, and as a consequence, links in the chain disappear from the target population over time. When the chain eventually runs out, so does the drive.[20] The built-in expiry date that daisy-chaining creates is useful for purposes of conducting safe field tests, as well as for ensuring that the wishes of particular nations are respected.[21] As noted in Chapter 3, ecosystems span state borders and thus, large-scale interventions will often affect multiple nations' environments. In the event that some nations aren't interested in having their environment affected by a particular gene drive, daisy-chaining could be used to let them opt out.[22]

A second worry is that genetic experiments must first be conducted in order to develop a genetic intervention, and that such experiments risk harming their subjects. As mentioned earlier, the harms risked by genetic experimentation are initial, rather than perpetual, and only a small subset of the target population are put at risk. Still, it's worth asking whether imposing those risks is justified. The main harms risked by genetic experiments are those associated with the unintended consequences an edit may have for a research subject. Consider a horror story from the 1980s. In 1985, the United States Department of Agriculture attempted to engineer pigs who would grow faster and thus, consume less grain before becoming fit for slaughter. The researchers' approach involved inserting a human growth hormone gene into a series of pig fetuses, 19 of which ended up surviving birth and reaching maturity. In all 19 cases, however, the genetic modification had unintended consequences. In particular, all of the pigs were sterile, and a number of them experienced medical issues such as physical deformity, ulcers, arthritis, and a compromised immune system.[23] Genetic engineering today is much more advanced than it was in the 1980s, but risks are still present. Unintended consequences can occur when an edit is 'off-target,' that is, when experimenters accidentally cut the wrong part of the gene, but even 'on-target' edits can produce unexpected physical consequences.[24]

In previous work, I've argued that the risks of genetic experiments are justified by the sheer scale of wild animal suffering (WAS). Though exposing animals to risk (not to mention the discomfort of being held captive in a laboratory) is regrettable, only a very strict deontologist should think that those risks aren't outweighed by the interests of the great many future animals – trillions, perhaps – who we could potentially save from a dismal fate.[25] There is one complication, however. In order for the potential benefits of gene drives to trump the risks of experimentation, we must have good reason to believe that those potential benefits will be actualized. If safe gene drives that effectively assist wild animals are, despite researchers' best efforts, unlikely to ever be developed, then the risks of experimentation are unjustified.[26] In light of this, I think it's important for researchers to undertake a certain amount of knowledge and capacity building before they move on to experiments that involve sentient animals. For example, researchers interested in developing gene drives that reduce r-strategist birth rates might begin by experimenting on r-strategists who probably aren't sentient, such as insects (it's worth noting that, as I mentioned above, successful gene drives have already been performed on laboratory mosquito populations).[27] Researchers might also wait until the political support necessary to implement gene drives hopefully develops, as well as contribute to the public dialogs necessary to create that support.

A noteworthy point about CRISPR is that it may have the power to ameliorate some of the ethical concerns it raises. As noted above, CRISPR has already been used, with some degree of success, to perform gene edits that promise to improve the welfare of farmed animals, for example, by creating resistance to tuberculosis and by preventing horn growth. Adam Shriver argues that we should take this sort of research a step further and attempt to engineer farm animals whose capacity to experience pain is reduced – a possibility that might well be pursued with respect to lab animals, as well.[28] In fact, a couple of past studies have already succeeded in creating genetically engineered mice who possess a lesser capacity for pain. In both cases, genetic modification succeeded (either wholly or partially) at inhibiting the affective dimension of pain, that is, the extent to which one is disturbed by or minds their pain, rather than one's sensory experience of it. Put another way, the mice were engineered to be incapable of *suffering*, but they were still able to experience *mere pain*.[29] Though the researchers performing these studies weren't specifically interested in benefiting animals, the Sculpting Evolution group at the MIT Media Lab is currently working on a project that aims to improve animal welfare. More specifically, they aim to develop CRISPR-modified lab mice who are suitable for use in experiments, but who won't experience suffering in the process. Their hope is to engineer a sort of trigger that, when activated, would completely block any suffering (without the use of pain killing drugs).[30]

Again, it's an open question whether using CRISPR to pursue welfarist goals is morally justified. The obvious worry is that such projects will contribute to the perception that animal exploitation is justified and thus, impede the reformist efforts of animal rights advocates. In the event that CRISPR-modified lab animals do become available, though, using them would at least be less morally costly than using animals who would experience significant suffering. Researchers might, for example, develop birth-rate-reducing gene drives in suffering-inhibited lab mice before attempting to develop those drives in wild rodents. In this way, suffering-inhibited lab animals could be incorporated into the capacity building phase. Furthermore, in the event that replicating suffering inhibition in other species is something that could be done easily and at little risk to research subjects, researchers might first produce a suffering-inhibited population of the relevant species before doing gene drive research on it, for example, produce a suffering-inhibited population of meadow voles who could then be used in experiments that aim to reduce meadow vole birth rates. In this way, suffering-inhibited lab animals might also be used in the final stages of research.

In the next section, I consider a variety of ways that gene drives might be used to reduce WAS. One of the possibilities I'll consider is whether modifications that inhibit suffering should be used not only to reduce the harms of experimentation, but to help wild r-strategists, too.

## 5.4 What To Do With Gene Drives?

Thus far I hope to have shown that gene drive research has the potential to develop morally acceptable, effective means of assisting wild animals on a large scale. I have not, however, said much about what sorts of gene drives may be worth researching. In previous work, I've argued that we should try to develop gene drives that would change r-strategists into K-strategists, that is, gene drives that would reduce r-strategist litter and clutch sizes, and increase the amount of energy r-strategist parents devote to each of their offspring.[31] In a similar vein, other authors have suggested that we attempt to change carnivores into herbivores.[32] Such proposals are obviously very ambitious and may turn out to be impossible. They would also cause various ecological side effects that would have to be offset by supplementary drives, for example, if we change carnivores into herbivores, then we should also be mindful of controlling the size of the populations our former carnivores preyed on. Considering the sheer scale of WAS, though, it's worth doing the research necessary to determine whether radical forms of behavior change are feasible and could be safely implemented. Furthermore, even if it turns out that behavior change is often too ecologically dangerous, there may still be some ecosystems in which implementing it would be safe, for example, ecosystems where predators play a smaller role and could be modified without causing too much disruption.[33]

In what remains of this chapter, I'm going to consider some alternatives to large-scale behavior change. In particular, I'll consider whether we should try to make certain species extinct,[34] and whether we should try to remove animals' capacity for suffering.[35] Though I'll argue that behavior change is ideal, I'll also argue that the above-mentioned alternatives are appropriate in some contexts.

Let's consider extinction first. Is there any merit in the proposal that we should make predators or r-strategists extinct? On the one hand, doing so would certainly prevent a great deal of harm from occurring. Extinction would prevent further acts of predation, and it would prevent doomed r-strategist infants from being brought into existence. What's more, so long as extinction is achieved by spreading sterility (a feat which, it turns out, gene drives can accomplish),[36] existing predators and r-strategists won't lose their lives (unlike when a habitat is destroyed).[37] On the other hand, there are a number of costs associated with causing extinction. First, sterilizing predators and r-strategists would deprive some existing animals – specifically K-strategist predators – of the opportunity to raise children. For animals who have an interest in forming a family and raising their young, becoming sterile is a significant loss. Second, widespread extinction would be ecologically risky – maybe even more so than widespread behavior change. Third, even if we could, perhaps with the use of supplementary drives, safely cause predators and r-strategists to become extinct, it seems clear to me that behavior change is preferable to extinction. My thought is a simple one – a world that contains many animals is better than a world that contains very few animals – so long as most of the animals live good lives. The main reason for this is that a world with many animals (all or most of whom flourish) contains greater total utility than a world containing fewer animals. Though promoting total utility to the exclusion of all else has counter-intuitive consequences – among other things, it leads us to the conclusion that we should produce large, unhappy populations instead of smaller, happy populations – it remains true that considerations of total utility carry some moral significance. As noted in Chapter 4, total utility matters a great deal when utility levels are negative, since a large population of beings whose lives aren't worth living is far worse than a small population of beings whose lives aren't worth living. But even when utility levels are positive, total utility matters – just not as much as average utility does.

In addition to possessing greater total utility, a world with more animal individuals and animal species is a more biodiverse one. The appeal of biodiversity is partially aesthetic, at least for those of us who enjoy the experience of observing different animals, or who even just like the thought that many different animals share the world with us. More substantially, though, biodiversity is important for stability, as ecosystems that contain higher levels of biodiversity are better able to withstand disruptions. After all, implementation is about more than just bringing something about. In addition to doing what's needed to create better states of affairs, we should also do what's needed to ensure that they remain in place.[38] In the event that we do manage to create ecosystems where most wild animals live flourishing lives, biodiversity would presumably have a role to play in ensuring that those ecosystems don't collapse.

Though behavior change is preferable to making sentient r-strategists and sentient predators extinct, we may wish to make certain other organisms extinct, namely organisms that cause harm to wild animals and which do not themselves possess morally significant interests, for example, harmful bacteria and parasites. A proposal along these lines was recently suggested by Kevin Esvelt, who argues that CRISPR can and should be used to wipe out the screw fly.[39] The screw fly – a nasty parasite that lays eggs in open wounds so that its larvae may eat the flesh of living mammalian hosts – was wiped out in North America a couple of decades ago. Though they were wiped out primarily in order to protect livestock, wild mammals benefited immensely, too. Unfortunately, the method used to exterminate them (the Sterile Insect Technique) is not effective in all areas of the world and thus, the screw fly still exists in South America and other places. CRISPR could be used to wipe these parasites out for good, and there doesn't seem to be any good reason not to do so.

A second alternative to behavior change is removing r-strategists,' and perhaps K-strategist prey species,' capacity to suffer. This proposal is premised on the above-mentioned distinction that some philosophers and cognitive scientists draw between the affective dimension of pain – the extent to which one is disturbed by or minds their pain – and one's sensory experience of it. Experiencing mere pain (pain without or with little of its affective dimension) would still function to help an organism avoid harmful stimuli, but it's also far less unpleasant than suffering. Removing r-strategists' capacity to suffer without removing their capacity to feel mere pain is similar to giving them a permanent pain killer. It promises to significantly reduce the unpleasantness of their lives without making them less well-adapted to their environment.

This proposal has one major virtue relative to extinction and behavior change – it has the potential to significantly improve r-strategists lives without significantly impacting the ecosystems they live in. More specifically, it would allow us to assist r-strategists without affecting their birth rates, reducing their evolutionary fitness, or interfering with the relevant predator–prey relationships.

Though the above proposal is admittedly quite promising, I do have a number of reservations about removing r-strategists' capacity to suffer. First, the capacity to suffer certainly makes some contribution to how well-adapted a species is. Though it's true that physical discomfort can motivate animals to avoid harmful stimuli, real suffering is far more memorable than physical discomfort. An animal who's capable of suffering is likely better at learning from her negative experiences, and at avoiding that which she identifies as the cause of negative experiences, than an animal who's incapable of suffering. I'm not just speculating – Shriver, in his discussion of suffering-inhibited lab mice, admits that pain's affective dimension is associated with some behavioral responses (which is how the researchers were able to tell that their experiments worked), and that the loss of those behavioral responses might make an animal less capable of learning to avoid dangerous parts of her environment.[40] Second, removing the capacity to suffer wouldn't suffice to make r-strategists' lives good – it would only make them less terrible. Third, it would be very surprising if removing the capacity to suffer didn't also reduce a being's capacity to have positive experiences. After all, deadening sensitivity, whether it be emotional or physical, does more than just reduce a being's capacity for negative experiences. People who use anti-anxiety medication, for example, often report that it deadens not only their ability to feel anxiety, but also their ability to feel emotions like humor or excitement.[41] But even if the capacity to suffer could be removed without deadening one's general sensitivity, I still think that being unable to suffer would negatively affect one's positive experiences. This is because our ability to appreciate positive experiences is likely, at least to some extent, contingent upon our having had negative experiences to compare them to, for example, excitement is pleasant in part because we know what boredom feels like, and joy is pleasant in part because we know what sadness feels like. As a result, being incapable of suffering would almost certainly reduce one's ability to experience pleasure.

It's worth noting that it may be possible to design gene drives that circumvent at least some of the above problems. For example, we might design gene drives that remove r-strategists' capacity to suffer, but only during the period of time when they're most likely to die a painful, premature death, for example, the first few weeks of their lives. Such gene drives would significantly benefit those who die a terrible death shortly after birth, but it wouldn't impede the evolutionary fitness of those lucky enough to live longer, nor would it impede their capacity to appreciate positive experiences. It is tempting to go even further and attempt to deprive young r-strategists of sentience altogether. That way, they wouldn't even experience mere pain. I suspect, however, that even the temporary removal of sentience would significantly impact r-strategists' adaptability. Merely removing the capacity to suffer is a much safer bet, even though it would mean that most r-strategists continue to die painfully and prematurely.

In light of the above, it seems clear to me that removing r-strategists' capacity to suffer is not an entirely adequate substitute for behavior change. That said, if it turns out that, even after considerable research, changing the behavior of r-strategists is infeasible, then removing their capacity to suffer is a second-best option that may be worth pursuing. Similarly, if behavior change does turn out to be feasible but too dangerous in some ecological contexts, removing r-strategists' capacity to suffer may be a desirable alternative in those contexts. In fact, if it turns out that we can limit the removal of suffering to r-strategists' early life, we may wish to conduct such gene drives alongside any behavior-change drives we implement. That way any r-strategists who, despite our best efforts, do die a painful and premature death, at least won't suffer in the process.

## 5.5 Conclusion

In conclusion, I've argued that gene editing technology provides a particularly promising means of assisting wild animals. Though gene drives do carry certain moral costs, those costs are considerably less than the costs of conventional wild life management. The reason is that, unlike conventional management, a gene drive's harms are merely initial, and they only affect a small subset of the target population.

In addition to arguing the merits of gene editing, I argued that it should ideally be used to change animals' dietary and reproductive behaviors, that is, to turn predators into herbivores and r-strategists in K-strategists. However, I admitted that gene drives should sometimes be used to eliminate species, particularly non-sentient parasites, and that gene drives should perhaps be used to remove r-strategists' capacity to suffer in the event that behavior change proves infeasible.

## Notes

1 Technically speaking, CRISPR just refers to a structural feature present in the genomes of different bacteria. The actual gene editing is done by an associated enzyme called Cas9 in combination with a guide RNA molecule that targets the desired part of the genome to be modified. Nevertheless, the acronym 'CRISPR' is now typically used to refer to this new form of gene editing. See Doudna and Charpentier (2014) p. 1 and p. 3. For an informative and accessible article about CRISPR, see Ledford (2015b).
2 The idea of dispersing a genetically engineered trait through gene drives has been around for a while. See Burt (2003).
3 Authors against intervention tend to assume that transforming nature into a zoo is the only available form of large-scale intervention. See Nussbaum (2007) p. 379, Donaldson and Kymlicka (2011) p. 164, and Donaldson and Kymlicka (2013) pp. 154–7.
4 For a discussion of this possibility, see Milburn (2019) p. 24 and Soryl (2019) pp. 63–4.
5 I originally encountered this idea when Andrew Fenton brought it up at a meeting of the Atlantic Regions Philosophers' Association. For a discussion of a similar proposal – the proposal that we should remove farm animals' capacity to suffer – see Shriver (2009) and Shriver and McConnachie (2018).
6 Pearce (2015), McMahon (2015), and Johannsen (2017).
7 Gersbach (2014) pp. 1009–10, Doudna and Charppentier (2014) p. 3, and Ledford (2015b) p. 21.
8 Ledford (2015a).
9 Zhang (2018).
10 See the company's website at www.the-odin.com/
11 For the types of gene drive that occur in nature, see Esvelt et al. (2014) pp. 2–3.
12 Ledford (2015b) p. 22 and Esvelt et al. (2014) pp. 3–4.
13 Esvelt et al. (2014) pp. 4–9. For studies that have successfully used CRISPR to conduct gene drives, see Gantz and Bier (2015), Gantz et al. (2015), and Hammond et al. (2016).
14 Gantz et al. (2015) and Hammond et al. (2016). For a general discussion of genetic methods for controlling mosquito populations, see Alphey (2014).
15 Gao et al. (2017).
16 For an accessible description of the relevant experiment, see Quinton (2019). For the academic article associated with the experiment, see Young et al. (2020).
17 Section 5.2 is a revised version of a discussion that was published in Johannsen (2017) pp. 340–1.
18 See endnote 3.
19 Esvelt and Gemmell (2017) and Noble et al. (2019).
20 Min at al. (2017) p. 2 and Esvelt and Gemmell (2017) p. 5.
21 Esvelt and Gemmell (2017).
22 It's worth noting that some work has been done on combining daisy-chaining with 'genetic quorums' for purposes of making gene drives reversible. For a preliminary paper on daisy quorum drives, see Min et al. (2017).
23 Comstock (1992) p. 196.
24 Schultz-Bergin (2018) pp. 227–31.
25 Johannsen (2017) p. 339 and p. 343.
26 Many thanks to Sue Donaldson for drawing my attention to this worry.
27 Gantz and Bier (2015), Gantz et al. (2015), and Hammond et al. (2016).
28 See Shriver (2009) and Shriver and McConnachie (2018).

29 For a helpful discussion of these studies, see Shriver (2009) p. 118. For the studies themselves, see Wei et al. (2002) and Sun et al. (2008). I also mentioned the distinction between suffering and mere pain in Chapter 2. Though I think the distinction is plausible, I suspect that the affective dimension is always present, at least to a small extent, in any experience we'd call 'pain.' See endnote 7 in Chapter 2 for some thoughts about this.

30 A description of the project is available at www.media.mit.edu/projects/reducing-suffering-in-laboratory-animals/overview/

31 Johannsen (2017).

32 Pearce (2015) and McMahan (2015).

33 For example, it's common for predators to play a smaller role in island ecosystems. For a discussion of how this affects prey species, behavior, see Cooper et al. (2014).

34 See endnote 4.

35 See endnote 5.

36 Hammond et al. (2016).

37 For a discussion of habitat destruction, see Chapter 4.

38 G.A. Cohen helpfully divides feasibility into two parts – accessibility and stability. See Cohen (2009) pp. 56–7.

39 Esvelt (2018).

40 Shriver (2009) p. 120.

41 See, for example, Read et al. (2014).

## References


Alphey, L. 2014. "Genetic Control of Mosquitoes." *Annual Review of Entomology* 59: 205–24.

Burt, A. 2003. "Site Specific Selfish-Genes as Tools for the Control and Genetic Engineering of Natural Populations." *Proceedings of the Royal Society of London* 270: 921–8.

Cohen, G.A. 2009. *Why Not Socialism?* (Princeton: Princeton University Press).

Comstock, G. 1992. "Should We Genetically Engineer Hogs?" *Between the Species* 8: 196–202.

Cooper, Jr., W.E., R.A. Pyron, and T. Garland Jr. 2014. "Island Tameness: Living on Islands Reduces Flight Initiation Distance." *Proceedings of the Royal Society B: Biological Sciences* 281: 20133019.

Donaldson, S., and W. Kymlicka. 2011. *Zoopolis: A Political Theory of Animal Rights* (New York: Oxford University Press).

Donaldson, S., and W. Kymlicka. 2013. "A Defense of Animal Citizens and Sovereigns." *Law, Ethics, and Philosophy* 1:143–60.

Doudna, J.A., and E. Charpentier. 2014. "The New Frontier of Genome Editing with CRISPR-Cas9." *Science* 346: 1258096.

Esvelt, K.M., A.L. Smidler, F. Catteruccia, and G.M. Church. 2014. "Concerning RNA-Guided Gene Drives for the Alteration of Wild Populations." *eLife* 3: e03401.

Esvelt, K.M., and N.J. Gemmell. 2017. "Conservation Demands Safe Gene Drive." *PLOS Biology* 15: e2003850.

Esvelt, K.M. 2018. "When Are We Obligated to Edit Wild Creatures?" Available at: www.leapsmag.com/when-are-we-obligated-to-edit-wild-creatures/ (Accessed on July 7th, 2020).

Gantz, V.M., N. Jasinskiene, O. Tatarenkova, A. Fazekas, V.M. Macias, E. Bier, and A.A. James. 2015. "Highly Efficient Cas9-Mediated Gene Drive for Population Modification of the Malaria Vector Mosquito *Anopheles Stephensi*." *PNAS* 112: E6736–43.

Gantz, V.M., and E. Bier. 2015. "The Mutagenic Chain Reaction: A Method for Converting Heterozygous to Homozygous Mutations." *Science* 348: 442–4.

Gao, Y., W. Haibo, W. Yongsheng, L. Xin, C. Linlin, L. Qian, C. Chenchen, L. Xu, J. Zhang, and Y. Zhang. 2017. "Single Cas9 Nickase Induced Generation of *NRAMP1* Knockin Cattle with Reduced Off-Target Effects." *Genome Biology* 18: 13.

Gersbach, C.A. 2014. "Genome Engineering: The Next Genomic Revolution." *Nature Methods* 11: 1009–11.

Hammond, A., R. Galizi, K. Kyrou, A. Simoni, C. Siniscalchi, D. Katsanos, M. Gribble, D. Baker, and E. Marois. 2016. "A CRISPR-Cas9 Gene Drive System Targeting Female Reproduction in the Malaria Mosquito Vector *Anopheles Gambiae*." *Nature Biotechnology* 34: 78–83.

Johannsen, K. 2017. "Animal Rights and the Problem of r-Strategists." *Ethical Theory and Moral Practice* 20: 333–45.

Ledford, H. 2015a. "Biohackers Gear Up for Genome Editing." *Nature* 524: 398–9.

Ledford, H. 2015b. "CRISPR, the Disruptor." *Nature* 522: 20–4.

McMahan, J. 2015. "The Moral Problem of Predation." In *Philosophy Comes to Dinner: Arguments about the Ethics of Eating*. Eds. A. Chignell, T. Cuneo, and M. Halteman (London: Routledge), pp. 268–94.

Milburn, J. 2019. "Sentientist Politics Gone Wild." In *Book Symposium: "Sentientist Politics: A Theory of Global Interspecies Justice*." By A. Cochrane, J. Milburn, and S. O'Sullivan, pp. 19–24 (*Politics and Animals*, Vol. 5).

Min, J., C. Noble, D. Najjar, and K.M. Esvelt. 2017. "Daisy Quorum Drives for the Genetic Restoration of Wild Populations." *BioRxiv* 115618.

MIT Media Lab. "Reducing Suffering in Laboratory Animals." Available at: www.media.mit.edu/projects/reducing-suffering-in-laboratory-animals/overview/ (Accessed on Feb 29th, 2020).

Noble, C., J. Min, J. Olejarz, J. Buchthal, A. Chavez, A.L. Smidler, E.A. DeBenedictis, G.M. Church, M.A. Nowak, and K.M. Esvelt. 2019. "Daisy-Chain Gene Drives for the Alteration of Local Populations." *Proceedings of the National Academy of Sciences* 116: 8275–82.

Nussbaum, M.C. 2007. *Frontiers of Justice* (Cambridge, MA: Belknap Press).

Pearce, D. 2015. "Reprogramming Predators." Available at: www.hedweb.com/abolitionist-project/reprogramming-predators.html (Accessed on Jan 1st, 2019).

Quinton, A. 2019. "Genome Edited Bull Passes on Hornless Trait to Calves: Study Sheds Light on Future of Genome Editing in Livestock." Available at: www.ucdavis.edu/news/genome-edited-bull-passes-hornless-trait-calves?fbclid=IwAR1mJAiqbd1C7pUOqvp5hvir8PQ3xJfPRzjpbd938I1oCmnHcSQ-QtwB2aM (Accessed on Oct 25th, 2019).

Read, J., C. Cartwright, and K. Gibson. 2014. "Adverse Emotional and Interpersonal Effects Reported by 1829 New Zealanders While Taking Antidepressants." *Psychiatry Research* 216: 67–73.

Schultz-Bergin, M. 2018. "Is CRISPR an Ethical Game Changer?" *Journal of Agricultural and Environmental Ethics* 31: 219–38.

Shriver, A. 2009. "Knocking Out Pain in Livestock: Can Technology Succeed Where Morality Has Stalled?" *Neuroethics* 2: 115–24.

Shriver, A., and E. McConnachie. 2018. "Genetically Modifying Livestock for Improved Welfare: A Path Forward." *Journal of Agricultural and Environmental Ethics* 31: 161–80.
Soryl, A.A. 2019. "Establishing the Moral Significance of Wild Animal Welfare and Considering Practical Methods of Intervention" (Master's Thesis, University of Amsterdam).
Sun, Y., Y. Gao, Z. Zhao, B. Huang, J. Yin, G. Taylor, and Z. Chen. 2008. "Involvement of P311 in the Affective, but not in the Sensory Component of Pain." *Molecular Pain* 4: 23.
The Odin. "Biotech Kits & Classes." Available at: www.the-odin.com/ (Accessed on Feb 29th, 2020).
Wei, F., C. Qiu, S. Kim, L. Muglia, J. Maas Jr., V. Pineda, H. Xu, Z. Chen, D. Storm, L.J. Muglia, and M. Zhuo. 2002. "Genetic Elimination of Behavioral Sensitization in Mice Lacking Calmodulin-Stimulated Adenylyl Cyclases." *Neuron* 36: 713–26.
Young, A.E., T.A. Mansour, B.R. McNabb, J.R. Owen, J.F. Trott, C.T. Brown, and A.L. Van Eenennaam. 2020. "Genomic and Phenotypic Analyses of Six Offspring of a Genome-Edited Hornless Bull." *Nature Biotechnology* 38: 225–32.
Zhang, S. 2018. "A Biohacker Regrets Publicly Injecting Himself With CRISPR." *The Atlantic*. Available at: www.theatlantic.com/science/archive/2018/02/biohacking-stunts-crispr/553511/ (Accessed on Feb 29th, 2020).

# 6

# INTERVENTION AND ANIMAL RIGHTS ADVOCACY

## 6.1 Introduction

If I'm right to claim that we have a collective, institutional obligation to assist wild animals on a large scale, then we must also have derivative obligations to fulfill at the individual level. Since large-scale assistance isn't currently being conducted, fulfilling our collective obligation to assist requires advocating for, supporting, and/or encouraging the implementation of a large-scale assistance program.

There are at least three issues concerning a personal obligation to advocate for assistance. One is the general issue of how much time and resources we're required to contribute to beneficence-related causes. In Section 6.2, I discuss Peter Singer's work on beneficence and argue that, in light of his argument, we have a fairly robust duty to contribute to beneficence-related causes.[1] However, I also argue that Singer assigns beneficence more weight than he should. He maintains that we're required to prevent something bad so long as we needn't sacrifice anything of 'comparable' significance, but I argue that the necessity of a 'significant' sacrifice suffices to block duties of beneficence.[2]

The second issue concerns how much time and resources we should devote to advocating for the assistance of wild animals, rather than to other causes. In part for simplicity's sake, I focus on comparing wild animal suffering (WAS) to traditional animal rights causes such as animal agriculture and animal experimentation. In Section 6.3, I argue that because WAS is such a neglected area, it should currently be prioritized by a limited number of advocates. However, I also argue that when we take what Pablo Gilabert calls a 'transitional standpoint,'[3] it becomes apparent that choosing between WAS and traditional animal rights causes is often unnecessary. Some courses of action indirectly address both WAS and traditional causes, and courses of action that directly address animal agriculture can indirectly address WAS.

The third issue concerns what means should be used to advocate for assistance. I focus specifically on whether illegal means are permissible. In Section 6.4, I argue that objectives which simultaneously address WAS and animal exploitation, may sometimes be permissibly pursued via illegal means. Whether illegal animal rights advocacy is ever permissible depends on whether the disagreement between animal rights advocates and animal welfarists is reasonable. Though I suggest that animal welfarism is not a reasonable view, and that illegal means are therefore permissible in principle, I also argue that considerations of strategy favor legal advocacy.

## 6.2 An Obligation to Contribute

The argument Peter Singer originally made in "Famine, Affluence and Morality,"[4] and has expressed in various other pieces since then,[5] is perhaps the best-known argument for the claim that we have a moral obligation to pursue beneficence-related causes. His article is about global poverty in particular, but the argument he makes there can, with only minor modifications, be applied to any moral cause that concerns the prevention of suffering and death. I've included an outline of his argument below:

> Premise 1: If we can prevent something bad without sacrificing anything of comparable significance, we ought to do it.
> Premise 2: Absolute poverty is bad.
> Premise 3: There is some absolute poverty we can prevent without sacrificing anything of comparable moral significance.
> Conclusion: We ought to prevent some absolute poverty [whatever amount we can without sacrificing anything of comparable moral significance].[6]

To both explicate and provide intuitive support for the first (and most controversial) premise, Singer discusses a hypothetical case where one sees a child drowning in a pond. On the assumption that one knows how to swim and that no one else is both in a position to and interested in rescuing the child, almost all of us would agree that one has a duty to perform the rescue. What's more, we would agree that the duty to rescue is unaffected by the necessity of certain sacrifices, for example, that one is required to rescue the child even if doing so will damage one's shoes, cell phone, hair-do, etc. According to Singer, the pond example demonstrates that beneficence is an important, rather weighty part of common-sense morality. More specifically, it demonstrates that if we can prevent something bad, such as the death of a child, without cost to ourselves or merely at the cost of goods less important than a child's life, such as one's shoes or cell phone, then we ought to do so. And since we all have the ability to alleviate at least some of the suffering and death caused by global poverty, without sacrificing anything of comparable moral significance, then we should do that, too.[7]

Though Singer's article focused on the importance of donating to so-called 'charities' (an inconvenient term, considering that he thinks contributing is obligatory rather than a matter of 'charity'), it's worth noting that this is merely an example of how one's beneficence-related obligations may be fulfilled. If some other course of action more effectively prevents poverty-related suffering and death, then we should take it, and if donating to charities is best combined with additional strategies, then we should do so. Singer's argument is best understood as an argument for the claim that we have a duty to engage in whatever activities most effectively prevent poverty-related suffering and death, whether it be lobbying governments, making donations, convincing others to make donations (advocating for donation), etc.[8]

Again, the reason we're interested in Singer's argument is that modified versions of it can be applied to any cause concerned with preventing suffering and death, so long as prevention is indeed possible and can be accomplished without sacrificing anything of comparable moral significance. And the requirements of his argument are quite strict. After all, most of the goods we (the world's affluent) enjoy are incomparable to the badness of suffering and death. Acting in accordance with Singer's conclusion thus requires sacrificing a great deal of personal resources, for example, time, income, etc., to the pursuit of beneficence-related causes.

For my part, I think Singer's right to claim that we have a duty to contribute to beneficence-related causes. However, I'm skeptical of the claim that only 'comparable' sacrifices are capable of blocking duties of beneficence. It's much more plausible to claim that one's required to prevent something bad so long as one can do so without sacrificing anything 'significant.' After all, the things one's likely to damage when rescuing a drowning child, for example, one's shoes, cell phone, and hair-do; are fairly insignificant. Such things are either non-essential, easily replaced, or both and thus, it's reasonable to expect one to give them up if necessary to save a child's life.[9] In contrast, if saving the child's life will foreseeably require that one spend a year in a hospital bed afterwards (suppose the potential rescuer is recovering from a serious medical condition and that physical exertion would significantly worsen it), it's not clear that the sacrifice is morally required. Nor is one obligated to donate a kidney, even when doing so would save another person's life.[10] Neither a kidney nor a year spent in a hospital bed are comparable to another person's life, but both are significant sacrifices. It thus seems more consistent with commonsense morality to assign beneficence less weight than Singer does.

Interestingly, Singer is aware that assigning beneficence the weight he does seems too demanding. In fact, he concedes that demanding standards can be counter-productive. His worry is that asking more than is psychologically manageable for most will discourage them from contributing at all. Singer's solution is to formulate a public standard of assistance that differs from what he thinks morality actually requires. More specifically, Singer suggests that advocates and the representatives of charities should only ask that people devote a relatively small percentage of their income to beneficence, and that praise and blame be doled out accordingly, that is, that those who donate the required percentage be praised for it, and that those who fail to do so be blamed for it.[11] Singer's public standard is basically a strategy for indirectly and partially implementing the requirements of morality, a strategy that's required in light of the need to mitigate the psychological effects of demandingness.

Singer's view that demandingness affects the public standard of assistance; but not the true weight of beneficence, is peculiar. Doesn't demandingness actually reduce the extent to which we're morally obligated to pursue beneficence? Upon reflection, I think that Singer's view of the relationship between demandingness and beneficence presupposes utilitarianism. To see this, consider Singer's view from an explicitly deontological perspective. Though deontology is a rather varied family of moral views, most versions of it accord a significant role to respect for autonomy. Monistic deontological views such as Immanuel Kant's attempt to reduce other moral duties to a more fundamental principle (or principles) of autonomy, and pluralistic deontological views treat respect for autonomy as one among a variety of moral considerations that must be taken into account when deciding what, all things considered, the required course of action is. Either way, though, from a deontological perspective, demandingness limits the actual requirements of morality. Claiming that one is morally required to prevent something bad so long as one needn't sacrifice something of comparable moral significance, is tantamount to claiming that we're required to subordinate our personal projects to the interests of others. Put another way, it treats one as if one is merely a means to the fulfillments of others' interests. Deontology's emphasis on reciprocal respect for autonomy thus limits the weight beneficence can plausibly be accorded. Whatever weight it's assigned must be compatible with a significant, personal prerogative to pursue one's own interests.

In addition, I think it's important to keep in mind that assigning excessive weight to beneficence crowds out more than just the pursuit of self-interest. It leaves insufficient space for non-beneficence values. Again, only from a utilitarian perspective does it make sense to claim that deference to other values is a matter of strategy. If we reject utilitarian monism and accept that there are fundamental moral values other than just beneficence, it follows that those values must be traded off against beneficence in cases of conflict, that is, that they affect the content of our all-things-considered obligations. And there are a great many values beneficence conflicts with. For example, we devote a portion of our resources to maintaining our social relationships: we buy gifts and do favors for our loved ones in order to express affection and loyalty. In so far as the resources we devote to these purposes could otherwise be used to prevent suffering and death, then valuing beneficence conflicts with valuing our relationships. We also devote resources to various other moral purposes, for example, to assist someone when we owe them a favor (assistance is a matter of reciprocity, in such cases) or to reward someone when we think they deserve it. Again, the pursuit of such values conflicts with beneficence in so far as the resources spent could be used to prevent suffering and death.[12] From a pluralistic perspective, other values aren't psychological road blocks that prevent us from doing what's right. Instead, they provide competing moral reasons that should be factored in when determining what qualifies as a significant sacrifice.

A final reason to believe Singer presupposes utilitarianism is that considerations of distributive justice militate, to an extent, against assigning beneficence the weight he does. More specifically, the view that only comparable sacrifices block duties of beneficence, implies that under conditions of non-compliance, one is required to do far more than one's share of the moral work, that is, that one is required to pick up the slack when other people aren't satisfying their beneficence-related obligations. But isn't this unfair? At first glance, it would seem that the world's preventable bad things should be divided equitably among the people capable of preventing them. Each of us is responsible for preventing our own share of what's bad, but not for preventing other people's shares, should others shirk their duty.[13] The opposite thought – that one's responsible not only for the consequences of one's own behavior, but for the consequences of others' behavior (including their omissions) – is a rather utilitarian one. Though deontologists should perhaps concede that others' moral failure generates a reason to do extra moral work, that reason is far weaker than the reason we have to do our own share of it. As such, I think that justice in the distribution of moral burden is best understood as an additional consideration in competition with beneficence. Though it's irresponsible to do no more than one's fair share – Singer rightly notes that one should save drowning children even if there are other adults standing around doing nothing – it's implausible to claim that one's required to go way beyond one's fair share, too.[14] Instead, the idea of a fair share supports my contention that 'significant,' and not just 'comparable' sacrifices, suffice to block duties of beneficence.

## 6.3 Wild Animal Suffering vs. Traditional Animal Rights Issues

In the previous section I accepted that, in light of Singer's argument in "Famine, Affluence and Morality," we have an obligation to pursue beneficence-related causes. However, I also argued that Singer assigns excessive weight to beneficence. Though beneficence is important, it can't plausibly exclude a prerogative to (within reasonable limits) prioritize one's own interests, nor should it always take precedence over the values it conflicts with. There's more to morality than beneficence, and in cases of conflict, a balance should be struck. I argued that substituting 'comparable' sacrifice with 'significant' sacrifice, strikes the right balance.

In this section I address the issue of how much priority WAS should receive. There are a great many beneficence-related causes. Global poverty, climate change, animal agriculture, and animal experimentation are also responsible for considerable amounts of suffering and death. Considering that there are many ways to exercise beneficence, what reason is there to shift resources toward WAS and away from other cause areas? And, for those interested in improving the lives of animals in particular, is there reason to focus on WAS rather than traditional animal rights issues? In part for the sake of simplicity, this section will focus on comparing WAS to other animal rights causes.

The most obvious reasons for prioritizing WAS were canvassed in Chapter 1. I noted that the scale of WAS is massive, and that it's a largely neglected issue. Recall that, because of the r-strategy, most individual, sentient wild animals don't flourish and may not even live lives worth living. Also recall that the number of sentient wild animals born into the world over a period of time, for example, 10 years, dwarfs the number of human beings or farm animals born. In light of the above, considerations of scale support prioritizing WAS over virtually any other issue. What's more, WAS has been largely neglected thus far. As a result, the marginal utility of devoting more resources to it is potentially quite high.

Chapter 1 also discussed one of the main challenges to WAS – the issue of whether it's tractable. We noted that presently the world lacks effective, large-scale means of assisting wild animals, and that the complexity of ecosystems makes it difficult predict what the effects of large-scale intervention would be. By comparison, animal experimentation and animal agriculture are simple problems. Solving them merely requires that we cease harmful practices and replace them with morally acceptable ones. This is easier said than done, of course. Mustering the political and social will to change widespread practices is challenging. What's more, many people have a vested interest in the continuation of those practices, for example, animal agriculture allows people to continue consuming the animal products they're accustomed to, and various parties' economic interests are tied to animal agriculture and animal experimentation. Still, some audiences are more sympathetic to the case against animal exploitation than to the case for large-scale intervention in nature. The main reason is the relationship each has with environmentalism. Whereas environmentalism is more or less neutral with respect to animal experimentation, and broadly supportive of either ending or reforming animal agriculture (animal agriculture contributes significantly to climate change),[15] many forms of environmentalism are opposed to large-scale interventions in nature.

Though the extent to which WAS is tractable is somewhat unclear, the arguments I made in previous chapters give us reason to be optimistic. In Chapter 4, I argued that the ecological risks of large-scale intervention shouldn't be paralyzing. For one, the fact that nature may contain net suffering suggests that unintended ecological damage wouldn't necessarily have negative consequences overall – its net effect could be positive. I also argued that any harms caused by intervention are also excusable, so long as they're unintended and weren't foreseeable. With respect to the manner in which we should intervene, I argued, in Chapter 5, that gene drives are especially promising. With enough research, we should be able to use them to safely assist wild animals on a large scale. Finally, if, in Chapter 2, I was right to argue that naturalness is not an intrinsically valuable property, then interfering with nature is not bad in and of itself. It's important to conduct the research and testing necessary to develop safe, effective interventions, but so long as we can intervene in a sufficiently safe manner, we ought to.

So long as we have reason to be optimistic about WAS's long-term tractability, it might seem fair to conclude that WAS should receive a considerably higher level of priority than traditional animal rights issues. However, an additional consideration favors traditional issues. As I've noted in a couple of places throughout the book, negative duties are typically weightier than positive duties. Though positive duties sometimes trump them when the stakes are quite high, negative duties normally have priority. In light of this, it's noteworthy that traditional animal rights issues concern the violation of negative duties. The suffering and death associated with animal agriculture and animal experimentation, is worrisome in part because we're causing it. Animal agriculture is particularly concerning in this respect, not only because it causes considerable harm to billions of domesticated animals, but also because it contributes significantly to climate change, thereby harming human beings and wild animals, too. In other words, traditional animal rights issues aren't just about preventing suffering and death, that is, they aren't just matters of beneficence. They're a matter of creating the social and institutional conditions necessary for all of us to better fulfill our duty not to cause harm.

It might seem a little peculiar to argue that negative duties support traditional animal rights advocacy. Though a duty to refrain from harming animals supports changing one's lifestyle so that one no longer contributes to harmful institutions, for example, it supports going vegan, it might seem that non-maleficence ends there. As most vegans are aware, though, thoroughly extricating one's self from animal exploitation is virtually impossible under present circumstances. Though refraining from ordering meat and dairy at restaurants is feasible enough, refraining from eating at (and thus supporting) restaurants that serve meat and dairy is very difficult. And though refraining from buying meat or dairy at one's grocery store is feasible enough, refraining from shopping at (and thus supporting) grocery stores that sell meat and dairy is very, very difficult. Perhaps more difficult than anything, though, is avoiding complicity in government support for animal exploitation. After all, governments support animal exploitation in various ways, whether it be through agricultural subsidies or through state-funded research involving animal experiments. As a result, we contribute to the perpetuation of animal exploitation simply by paying taxes. Of course, none of these considerations suggest that it's pointless to refrain from eating meat and dairy, from buying clothes that contain animal-based materials, etc. Such efforts don't eliminate one's contribution to animal exploitation, but they do reduce it. Still, the fact that it's virtually impossible to extricate one's self from animal exploitation under present circumstances gives us extra reason to count animal agriculture and animal experimentation, among our cause areas. Again, pursuing these areas contributes to creating the social and institutional conditions needed for all of us to better fulfill our negative duties.

The considerations discussed so far suggest a stalemate between WAS and traditional animal rights issues. On the one hand, WAS is much larger in scale and is relatively neglected. On the other hand, traditional issues may be somewhat more tractable, and pursuing them is supported by not only beneficence, but also by our negative obligation to refrain from causing harm. Thankfully, there's less tension between these cause areas than there appears to be. I argue that when we take what Pablo Gilabert calls a 'transitional standpoint,'[16] we see that some courses of action simultaneously address both WAS and traditional cause areas, and that directly addressing animal agriculture can indirectly address WAS. I'll begin by describing the idea of a transitional standpoint, after which I'll describe how Gilabert's thoughts help ease the tension between addressing WAS and addressing traditional cause areas.

Gilabert's ideas about transitioning to morally superior states of affairs revolve around a distinction between hard and soft feasibility conditions, and the implications of this distinction for our moral obligations.[17] Though the claim 'ought implies can' is true in a sense, the relationship between feasibility and obligation is more complicated than it appears. Hard feasibility conditions, on the one hand, straightforwardly constrain our obligations. If a particular state of affairs is logically impossible, or it's physically impossible for any person or group of people to implement, then implementation cannot be obligatory. Soft feasibility conditions, on the other hand, are less straightforward. Soft feasibility constrains what we're able to accomplish here and now, but they aren't necessarily permanent. What makes a feasibility condition 'soft' is that it can be either altered or removed and thus, a state of affairs that's infeasible in the soft sense can be made to become feasible in the future. According to Gilabert, this implies that some of our moral obligations are dynamic. Though certain desirable states of affairs are currently infeasible, their desirability can generate an obligation to bring about other, feasible states of affairs from which they're accessible.[18] Soft feasibility conditions include factors like the current state of public opinion, or the economic and political institutions currently in place.

From a transitional perspective, it's clear that pursuing WAS doesn't always conflict with pursuing traditional animal rights issues, and vice versa. For example, philosophical arguments that defend animals' moral status indirectly contribute to all of these areas. Drawing attention to the fact that we treat all sentient human beings equally, regardless of their other cognitive capacities, is sometimes effective in persuading people that, as a matter of consistency, we should extend equal consideration to sentient animals.[19] And once one accepts that moral equality extends to sentient animals, one is logically committed to rejecting behavior that's inconsistent with equal consideration. Of course, a person's behavior can be inconsistent with their beliefs, and the adoption of a new belief doesn't necessarily mean that they'll revise their other beliefs in light of it. Convincing someone of animals' morally equal status is thus no guarantee that they'll, for example, stop eating meat, or that they'll even adopt the view that they ought to stop eating meat. Still, cognitive dissonance is uncomfortable. Arguments for moral status can thus serve as a philosophical stepping stone to other changes in an audience's beliefs and behaviors.

Accepting that moral equality extends to animals straightforwardly implies that one should reject animal agriculture and animal experimentation (as least as they're currently practiced), but it also implies that one should think critically about suffering in the wild and be open to the possibility that it should be prevented. After all, animals have an interest in avoiding suffering, and if one takes the interests of animals seriously, then the fact that suffering in the wild is so extreme and pervasive should be concerning. In so far as the arguments for extending moral equality to animals tends to fall on deaf ears, though, an alternative is to focus on how conventional attitudes toward animals are often inconsistent with each other.[20] For example, many people care a great deal about the interests of pet animals – not just their own pets (if they even have pets), but pet animals more generally. As a result, the thought of dogs, cats, etc., being eaten, abused, or left unassisted, is much more disturbing to them than the thought of farm animals or wild animals, being eaten, abused, or left unassisted. Drawing attention to the inconsistency in according such a high level of significance to pet animals' interests, and yet so little significance to the interests of other animals, can sometimes be an effective way of convincing an audience to extend the concern they have for pet animals to other animals.

Another course of action that (indirectly) addresses both WAS and traditional animal rights causes, is advocacy for and argumentation in support of representing animals' interests in democratic processes.[21] Formally representing animals' interests in, for example, not being killed, not being made to suffer, and in receiving aid when in need would make it more feasible to reform exploitive practices and implement a large-scale assistance program. There are various complications of course – we might wonder which animals should have their interests represented, which levels of government they should be represented at, the manner in which their interests should be represented, etc. Since animals can't literally cast a vote or hold office, the political expression of their interests would have to be mediated by human representatives. Regardless of the details, though, it's clear that democratic reforms would be useful for securing other measures that must be collectively implemented. And so long as we take an inclusive approach to representation that counts not only domesticated animals, but also wild animals, then we can expect that democratic reforms will benefit both groups. That said, if we do include wild animals in democratic processes, we should likely represent their interests differently than the interests of domesticated animals, considering that we have a much closer relationship with the latter. For example, wild animals' representatives could be given the right to participate in political discourse, but not to vote on legislation, whereas domesticated animals' representatives would have the right to both engage in discourse and vote. Such an arrangement seems appropriate in light of the fact that wild animals reside in our jurisdictions and are affected by our political and economic decisions, but unlike domesticated animals, are uninvolved (or are at least much less involved) in social cooperation with us.

An additional point worth noting is that some of the social practices that directly support animal agriculture indirectly support taking a dismissive attitude toward WAS. For example, consider the process via which children are socialized into eating meat and dairy.[22] Parents who are themselves in the habit of eating meat and dairy typically serve it to children at a young age, and eating animal products thus becomes habitual long before one has the ability to make an autonomous choice. Interestingly, though, the process of being socialized into eating animal products often has some hitches. When children find out that, for example, 'chicken' is literally dead chicken flesh, they often have a visceral reaction to the news and sometimes refuse to eat meat afterwards. Their rebellion is often quashed, though, when meat continues to be served to them, when they're pressured to eat it and threatened with penalties (No pudding until you eat your meat!), and when it's explained to them that the beings who were killed to produce it are 'just animals.'[23] Parents aren't the only parties who take part in this socialization process, of course. Teachers, priests, advertisements, and a variety of other factors contribute too. The point is just that the social processes that support consuming animal products function to habituate consumers at a very young age, smother their sympathy for the animals who are killed, and persuade them to believe that animals matter less. Now suppose that this socialization process was to be significantly disrupted, perhaps by convincing a sufficiently large number of recent and future parents to become vegan. One obvious result would be that fewer children are socialized into consuming animal products. Another is that more people, as they grow up, would decide that they're actively opposed to animal agriculture. I expect that the effects would be more widespread, though. As we noted above, one's general view concerning the moral status of animals has various implications for one's beliefs and behaviors. Since socializing someone to consume animal products involves, in part, socializing them to believe that animals matter less, and socializing them to suppress their sympathy for the animals killed, presumably disrupting that process would serve to both promote sympathy for animals and disrupt the prevalence of the view that animals are morally inferior. The end result would be more people in the world with attitudes and beliefs that are conducive to caring about WAS.

As we've seen, the apparent tension between addressing WAS and addressing traditional animal rights issues dissolves somewhat when we take a transitional standpoint. Some courses of action serve to indirectly address both WAS and traditional issues, and directly challenging animal agriculture serves to indirectly address WAS. Notwithstanding the above, though, I do think it's important that some advocates work directly on WAS. As noted earlier, WAS is a largely neglected issue, and so the marginal utility associated with devoting one's self to it is potentially quite high right now. More concretely, whatever efforts we devote right now to raising awareness about WAS, researching the different sources of suffering in nature, and determining which interventions are the most promising, will be useful once the social and political conditions conducive to implementing those interventions are in place. One thing to note, though, is that from a transitional standpoint, some forms of direct advocacy make more sense than others. For example, many people find the idea of intervening in natural processes such as predation or the r-strategy to be counter-intuitive. Though I hope the arguments I've presented in previous chapters suffice to show that, under the right conditions, intervening in these processes is justified (morally required, in fact), it would be naïve to ignore people's gut reaction. Instead of advocating for the end of predation or the end of the r-strategy, it makes sense to, at the moment, pursue more modest objectives. Some promising options include raising awareness about the extent and severity of suffering in the wild, presenting arguments for why we should care about that suffering, and promoting some of the more promising small and medium-scale interventions. For example, I think that many people find it easy to get on board with interventions designed to prevent disease and/or to kill non-sentient parasites. Audiences need only take a small step to conclude that if we're already administering vaccines to wild animals,[24] or wiping out screw fly populations,[25] for anthropocentric reasons, then perhaps we should do more of that sort of thing but for wild animals' sakes. Once enough awareness about WAS has been raised and there's widespread support for various small and medium-scale interventions, perhaps the public will be more receptive to the claim that we should be researching more dramatic interventions.

## 6.4 Legal Advocacy vs. Illegal Advocacy

In the previous section, I discussed a variety of objectives advocates should consider pursuing. Some of those objectives don't address WAS directly, but addressing it indirectly is also worthwhile. In this section, I'm concerned with the means via which both direct and indirect objectives should be pursued. More specifically, I'm concerned with whether it's permissible to use illegal means, for example, violence, property destruction, etc.

At first glance, the answer to whether illegal means of addressing WAS are permissible is a straightforward 'no.' To see why, recall our comparison of distributive justice with beneficence in Chapter 3. There, we noted that one of the main virtues of using beneficence, rather than distributive justice, to ground interventions, is that beneficence is neutral between the animal rights position and animal welfarism. Though I myself hold the animal rights position and consider it to be worth advocating for, arguments for intervention should avoid presupposing the animal rights position so long as it remains a minority view. Building widespread support for intervention requires offering reasons that others can accept, and until the public accepts that moral equality extends to animals (an event that may or may not occur one day), appeals to beneficence remain more effective than appeals to distributive justice. What's more, I noted that our interest in building support is about legitimacy, not just strategy. The reason is that state-provided assistance for wild animals must be democratically enacted if it's to be legitimate. Though fulfilling our collective *negative duties* to *domesticated animals* is arguably a requirement of legitimacy, fulfilling our collective *positive duties* to *wild animals* is not. Failing to satisfy positive duties to wild animals is akin to failing to provide an adequate level of assistance to impoverished nations. Affluent countries that fail to provide adequate assistance are blameworthy, but they aren't illegitimate, and assisting legitimately requires democratic enactment. In contrast, animal agriculture is akin to slavery, and slavery is illegitimate regardless of whether it enjoys the majority's support. Though it's plausible to maintain that animal agriculture undermines legitimacy, the same isn't true of failing to assist wild animals.

If failing to assist wild animals does not compromise legitimacy, then it would seem that pursuing it via illegal advocacy is unjustified. After all, illegal advocacy, at least when justified, is normally a response to illegitimacy. Those under the jurisdiction of a legitimate authority have an obligation to obey its laws precisely because it's a legitimate authority. When an authority loses its legitimacy, the requirement to obey its laws is also lost and illegal advocacy becomes appropriate (though some laws should be obeyed for reasons other than legitimacy, for example, laws that enforce pre-existing moral obligations such as the obligation not to kill and laws necessary for social coordination). Of course, illegal advocacy is often still appropriate under legitimate regimes. Justifying civil disobedience, for example, does not require that the state be tyrannical. Only more extreme forms of advocacy, such as terrorism, require that the state be illegitimate. What civil disobedience requires is localized illegitimacy, such as a particular illegitimate law within an otherwise legitimate legal system.[26] Regardless, the appropriateness of illegal advocacy requires illegitimacy, and if refraining from aiding wild animals is legitimate (though wrong), then it seemingly can't justify illegal advocacy.

There's a difficulty with the above argument, though. In the previous section, I noted that a number of objectives addressing traditional animal rights issues also address WAS. In light of this connection, perhaps illegal means are permissible so long as they also address traditional issues, for example, using civil disobedience to promote veganism in hopes of disrupting the social processes that both suppress children's sympathy and instill the belief that animals matter less.

Illegal advocacy that challenges animal exploitation might be justified in some circumstances. I think that the answer depends on whether disagreement about animals' moral status is part of, or prior to, reasonable pluralism. If it's true that animals have basic moral rights such as a right to life, and that those rights constrain the scope of democratic authority, then a state's failure to protect those rights undermines its legitimacy.[27] In contrast, if the animal rights position and animal welfarism are both reasonable views and thus, the issue of whether animals have rights is a matter of reasonable disagreement, then a state's failure to protect animals' interests does not undermine legitimacy, and legislation that seeks to protect those interests would require democratic support to be legitimate.[28] After all, if animal welfarism is reasonable, then laws which, for example, prohibit the killing of animals, must be justified using reasons that are acceptable from a welfarist perspective. Since appealing to animals' right to life is not acceptable from a welfarist perspective, an appeal to majority rule would be needed instead.[29]

Whether the disagreement between animal welfarists and those who hold the animal rights position is a part of reasonable pluralism, is a difficult issue that I don't hope to settle here. However, I will take this opportunity to express some doubts about the claim that animal welfarism is reasonable. Defenders of the animal rights position often claim that animal welfarism is prejudiced – specifically speciesist.[30] They maintain that viewing animals' interests as less worthy of consideration is analogous to viewing women's interests, non-white people's interests, and the interests of people with disabilities, as less worthy of consideration. In other words, they maintain that it's analogous to sexism, racism, and ableism. By asserting this, defenders of the animal rights view aren't just saying that animal welfarism is incorrect – they're claiming that it's unworthy of respect. After all, a prejudiced view can't also be a reasonable view. The disagreement between those who maintain that all humans are equal, and those who hold various prejudices, is not on par with, for example, political disagreement between conservatives and socialists.

Determining whether a view is worthy of respect can sometimes be difficult, but we do have some criteria to use.[31] In particular, the reasons given in support of a reasonable view should meet certain basic standards of argumentation. Patently false empirical claims are certainly a red flag, for example, anti-Semitic claims that 'Jews' faked the holocaust, that 'Jews' control the word, etc. Another red flag is appeals to popularity, for example, 'Everyone knows that it's okay to kill farm animals!.' Yet another red flag is when the reasons given in favor of a view are applied inconsistently. For example, were one to claim that killing cows, pigs, etc., is permissible because animals matter less, then one should also believe that it's permissible to kill other animals. As we noted above, though, animal welfarists normally maintain that it's impermissible to kill pet animals such as dogs and cats. Similarly, were one to claim that animals matter less because of some cognitive inability, for example, the inability to speak a language or the inability to engage in moral reasoning, then one should believe that human beings who lack the relevant ability matter less too. As we know, though, animal welfarists rarely maintain that human beings with severe cognitive disabilities matter less.

Again, I don't hope to definitively prove that animal welfarism is a prejudice. However, the fact that animal welfarism is typically accompanied by such inconsistent reasoning, is at least a good reason to think that it's a prejudice. And if animal welfarism is just a popular prejudice – one that we're socialized into from childhood because it supports institutions such as animal agriculture – then we needn't worry about whether appeals to animal rights are acceptable from a welfarist perspective any more than we need worry whether appeals to human rights are acceptable from a sexist or racist perspective. The justification given for a law need only be acceptable from reasonable points of view, not prejudiced points of view.

At first glance there might seem to be some tension between the view that animal welfarism is unreasonable, and the view that beneficence's neutrality is a virtue. If animal welfarism is prejudiced, then why does the legitimacy of intervening in nature require appealing to reasons that welfarists can accept? The answer is that positive rights are a matter of reasonable disagreement. Though all reasonable points of view agree on the importance of basic negative rights, such as a right not to be killed or a right not to be arbitrarily imprisoned, there's considerable disagreement about positive rights such as a right to financial assistance in times of need. While left-leaning political views often assert positive rights, conservative views often reject them. What's more, while some nationalist views allow that we have a collective duty to assist co-citizens, they deny that we owe any such duty in the absence of a special relationship. In light of such disagreements, the claim that we have a collective duty to assist wild animals is not something that can legitimize state action on its own. Positive rights, such as a right to assistance, are too contentious, and our relationship with wild animals if far less close than our relationship with domesticated animals. In other words, it's not that I think animal welfarism is a view worthy of respect. It's that legitimate assistance requires democratic support, and the best way to build that support is by appealing to reasons that are neutral between animal welfarism and the animal rights position.

If I'm right to claim that animal welfarism is unreasonable, then illegal advocacy that challenges animal exploitation is justifiable in principle. However, just because it's justifiable in principle doesn't mean that it's justified under normal circumstances. Illegal advocacy is only justified when it's more effective than legal advocacy, and I doubt that it normally is. Illegal acts such as violence and property destruction tend to alienate the public, thus discouraging support for reform. Less dramatic forms of illegal animal rights advocacy are also unlikely to attract much public support, given the prevalence of animal welfarism. Legal advocacy thus seems to make more sense strategically.

## 6.5 Conclusion

In conclusion, this chapter has discussed three questions concerning a personal obligation to advocate for assistance: (a) to what extent are we required to contribute to beneficence-related causes; (b) should WAS be prioritized over traditional animal rights causes; and (c) is it permissible to perform WAS advocacy through illegal means? With respect to (a), I argued that although we have an obligation to contribute to beneficence-related causes, we aren't morally required to make significant sacrifices. With respect to (b), I argued that although direct work on WAS should be prioritized by some advocates, choosing between WAS and traditional animal rights causes is often unnecessary, as it's frequently possible to simultaneously address both. With respect to (c), I suggested that it's permissible in principle to pursue objectives that simultaneously address WAS and animal exploitation. However, I also argued that illegal advocacy is not normally justified in practice, as legal advocacy is strategically superior.

## Notes

1 Singer (1972).
2 I have in mind the weaker principle that Singer mentions but does not endorse. See Singer (1972) p. 231 and p. 241.
3 Gilabert has been discussing the idea of a 'transitional standpoint' for over a decade now across various works. For a recent discussion, see Gilabert (2018) chapter 3.
4 Singer (1972).
5 For example, the argument also appears, with modifications, in Singer (1979); and in Singer (2010).
6 Singer (1979) p. 148.
7 Singer (1979) pp. 147–8.
8 Singer (1972) pp. 239–40.
9 In fact, Singer originally presented the pond case as an application of the more moderate principle I defend here, i.e., the principle that "if it is in our power to prevent something very bad from happening, without thereby sacrificing anything morally significant, we ought, morally, to do it." See Singer (1972) p. 231. It's only in later work that the pond case is used to illustrate the stronger, 'comparable sacrifice' principle that Singer himself defends. See Singer (1979) p. 147.
10 Arthur (1996) p. 157.
11 Singer (2010) chapter 10.
12 For a discussion of some of the values one has a prerogative to pursue, see Estlund (1998) pp. 101–7. Estlund's article is a response to G.A. Cohen's claim that principles of egalitarian justice directly apply not only to institutions, but to our personal choices. Estlund's discussion is still relevant to the demands of beneficence, though.
13 Singer discusses the 'fair shares' objection in Singer (2010) pp. 140–6. He's specifically responding to Liam Murphy and Kwame Anthony Appiah. See Murphy (2000) p. 76 and Appiah (2006) pp. 164–5.
14 Singer (2010) pp. 144–6.
15 For a discussion of animal agriculture's contribution to climate change, see Schlottmann and Sebo (2019) pp. 116–17.
16 Gilabert (2018) p. 81.
17 Gilabert (2018) pp. 66–9.
18 Gilabert (2018) pp. 79–80.
19 This sort of argument is often called 'the argument from marginal cases.' For an extended version of it, see Singer (1975) chapter 1.
20 See O'Sullivan (2016) pp. 57–9 and pp. 63–5.
21 For a discussion on representing animals' interests in democratic processes see, for example, Donaldson and Kymlicka (2011) chapter 5, Garner (2016), Parry (2016), and Cochrane (2018) chapter 3.
22 For a discussion of this process, see Singer (1975) pp. 224–7.

23 Of course, not all meat-eating parents pressure their children to eat meat, and there are instances where children successfully boycott meat despite having parents who eat it. For a relevant study that focuses on Australian families, see Bray et al. (2016).
24 See, for example, Buddle et al. (2011) and Rupprecht et al. (2003).
25 Vargas-Terán et al. (2005).
26 See Rawls's discussion of the difference between civil disobedience and military resistance, in Rawls (1971) pp. 363–8.
27 For a defense of the view that respect for animal rights is necessary for legitimacy, and that illegal advocacy is therefore justifiable in principle, see Cochrane (2018) chapter 6 and pp. 135–8.
28 This sort of view is defended in Basl and Schouten (2018).
29 Though I don't discuss it, some would argue that laws protecting animals' interests are illegitimate even if supported by the majority, as such laws arguably create impermissible restrictions on citizens' ability to pursue their conception of the good. For a discussion, see Berkey (2017) pp. 688–9.
30 This is one of Singer's main claims in Singer (1975).
31 See the criteria discussed in Dworkin (1977).

## References

Appiah, K.A. 2006. *Cosmopolitanism* (New York: Norton).

Arthur, J. 1996. "Famine Relief and the Ideal Moral Code." Reprinted in *Ethical Issues: Perspectives for Canadian, 3rd Edition* (Peterborough: Broadview Press, 2009), pp. 155–68.

Basl, J., and G. Schouten. 2018. "Can We Use Social Policy to Enhance Compliance with Moral Obligations to Animals?" *Ethical Theory and Moral Practice* 21: 629–47.

Berkey, B. 2017. "Prospects for an Inclusive Theory of Justice: The Case of Non-Human Animals." *Journal of Applied Philosophy* 34: 679–95.

Bray, H.J., S.C. Zambrano, A. Chur-Hansen, and R.A. Ankeny. 2016. "Not Appropriate Dinner Table Conversation? Talking to Children about Meat Production." *Appetite* 100: 1–9.

Buddle B.M., D.N. Wedlock, M. Denis, H.M. Vordermeier, and R.G. Hewinson. 2011. "Update on Vaccination of Cattle and Wildlife Populations against Tuberculosis." *Veterinary Microbiology* 151: 14–22.

Cochrane, A. 2018. *Sentientist Politics: A Theory of Global Inter-Species Justice* (Oxford: Oxford University Press).

Donaldson, S., and W. Kymlicka. 2011. *Zoopolis: A Political Theory of Animal Rights* (New York: Oxford University Press).

Dworkin, R. 1977. "Liberty and Moralism." In *Taking Rights Seriously*. By R. Dworkin (Cambridge, MA: Harvard University Press), pp. 240–58.

Estlund, D. 1998. "Liberalism, Equality, and Fraternity in Cohen's Critique of Rawls." *The Journal of Political Philosophy* 6: 99–112.

Garner, R. 2016. "Animals, Politics and Democracy." In *The Political Turn in Animal Ethics*. Eds. R. Garner and S. O'Sullivan (London: Rowman & Littlefield), pp. 103–17.

Gilabert, P. 2018. *Human Dignity and Human Rights* (New York: Oxford University Press).

Murphy, L. 2000. *Moral Demands in Nonideal Theory* (New York: Oxford University Press).

O'Sullivan, S. 2016. "Animals and the Politics of Equality." In *The Political Turn in Animal Ethics*. Eds. R. Garner and S. O'Sullivan (London: Rowman & Littlefield), pp. 51–67.

Parry, L.J. 2016. "Deliberative Democracy and Animals: Not So Strange Bedfellows." In *The Political Turn in Animal Ethics*. Eds. R. Garner and S. O'Sullivan (London: Rowman & Littlefield), pp. 137–53.

Rawls, J. 1971. *A Theory of Justice* (Cambridge, MA: Belknap Press).

Rupprecht C.E., C.A. Hanlon, and D. Slate. 2003. "Oral Vaccination of Wildlife Against Rabies: Opportunities and Challenges in Prevention and Control." *Developments in Biologicals* 119: 173–84.

Schlottmann, C., and J. Sebo. 2019. *Food, Animals, and the Environment* (New York: Routledge).

Singer, P. 1972. "Famine, Affluence, and Morality." *Philosophy & Public Affairs* 1: 229–43.

Singer, P. 1975. *Animal Liberation* (New York: Avon Books).

Singer, P. 1979. "Rich and Poor." In *Practical Ethics*, By P. Singer (Cambridge: Cambridge University Press), pp. 158–81. Reprinted in *Ethical Issues: Perspectives for Canadians, 3rd Edition*. Ed. E. Soifer (Peterborough: Broadview Press, 2009), pp. 141–54.

Singer, P. 2010. *The Life You Can Save* (New York: Random House).

Vargas-Terán M., H.C. Hofmann, and N.E. Tweddle. 2005. "Impact of Screwworm Eradication Programmes Using the Sterile Insect Technique." In *Sterile Insect Technique*. Eds. V.A. Dyck, J. Hendrichs, and A. Robinson (Dordrecht: Springer).

# INDEX

For Product Safety Concerns and Information please contact our EU representative GPSR@taylorandfrancis.com
Taylor & Francis Verlag GmbH, Kaufingerstraße 24, 80331 München, Germany

www.ingramcontent.com/pod-product-compliance
Ingram Content Group UK Ltd.
Pitfield, Milton Keynes, MK11 3LW, UK
UKHW022200190926
475702UK00003B/132
*9780367275686*